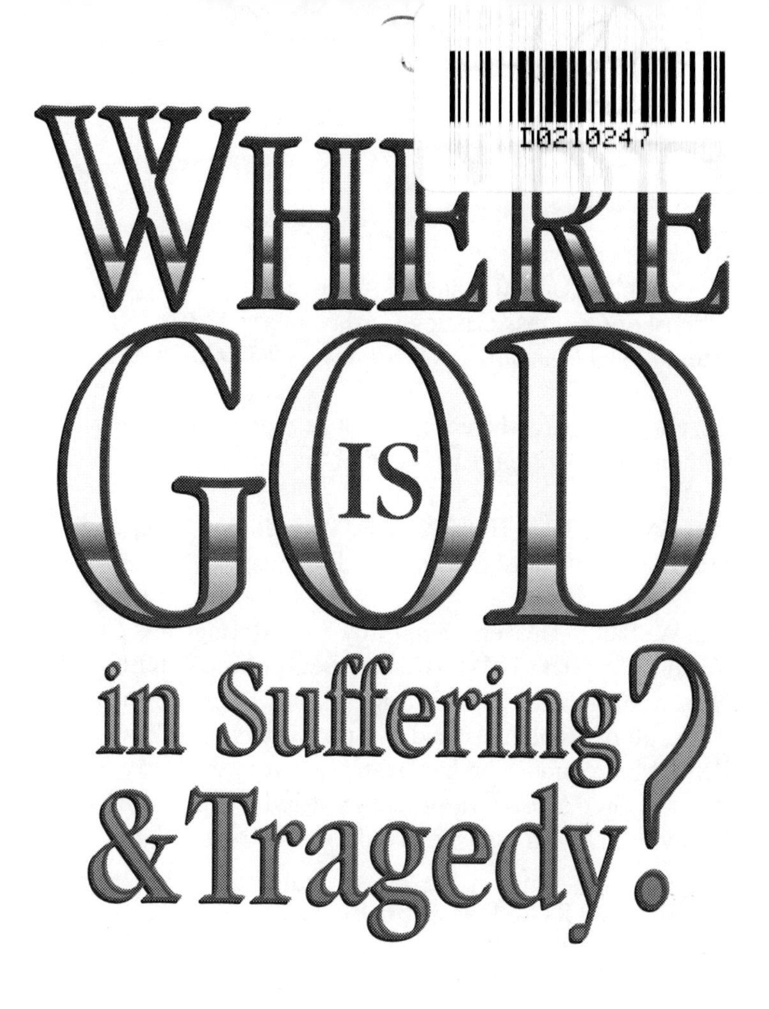

Thaddeus Barnum

Printed in the United States of America
Library of Congress Catalog Card Number: 97-076055
International Standard Book Number: 1-883928-28-1

Dedication

For Erilynne

She opens her mouth with skillful and godly Wisdom,
And in her tongue is the law of kindness—giving
counsel and instruction...
Her children rise up [Kris, Susan, Jill, Jan; women of
excellence in Jesus] and call her blessed;
And her husband boasts of and praises her, saying,
Many daughters have done virtuously, nobly and well
[with the strength of character that is steadfast in goodness]
But you excel them all.
Taken from Proverbs 31:26, 28-29 (AMP)

Previous book by Thaddeus and Erilynne Barnum: *Remember Eve*

To contact the author:

The Rev. Thaddeus Barnum
The Mustard Seed Project
19 West Elm Street
Darien, CT 06820
(203) 655-8446

Published by:
Longwood Communications
397 Kingslake Drive
DeBary, FL 32713

Acknowledgments

The actual facts surrounding Flight 427 recorded in this book are as exact as possible. I ask the readers to forgive any unintentional omissions. I want to thank the people mentioned in the book who kindly gave me permission to use their names. Some people I could not find, so I changed their names to protect them and their families. I also ask the reader to understand the nature of the wrestlings and questions of this book. It is written by someone in suffering, and written to those who know suffering and are asking the same questions.

I want to remember the crew and passengers of Flight 427, their family members, the rescue workers, the Salvation Army, Wayne Tatalovich and his family, the untold champions who gave themselves sacrificially to the rescue operation in Hopewell Township, and to our Christian family at Prince of Peace Church.

To God be the glory,
Thaddeus Barnum

Foreword

Inevitably there comes to each of us a cataclysm of nature, a horrible crime, a huge accident—an event that grips us and rudely shakes our complacent security.

Such was the crash of Flight 427.

I accompanied Thad Barnum to the crash site. Ours was the fundamental ministry of the church: to be present with people in crisis, to go where they had to go and to see what they had to see, to support them in the soul-wrenching work they had to do.

We were not technical people. We were ministers who had come to be with those who were at the edge of their emotional and spiritual resources. We came to offer what we had—faith in God, who because His own dear Son was broken on a hillside, was surely present in the midst of the carnage and desolation of that hillside.

As horrible as this crash was, we saw thousands of people respond with tireless and sacrificial love. We saw our community come together in a truly remarkable way. But most important, we saw faith in God wonderfully quickened.

From his own experience of this public tragedy, Thad Barnum beautifully and sensitively tells the story of being chaplain at the crash site of Flight 427. He describes the shock and horror that shakes our very foundations. He raises all the deep questions that challenge our basic security. He testifies to the wonderful and mysterious way of God's abiding love. His book is an encouragement and inspiration to all who wonder, "How can such things happen? Where is God?"

The Rt. Rev. Alden M. Hathaway
Bishop of Pittsburgh, Retired

Contents

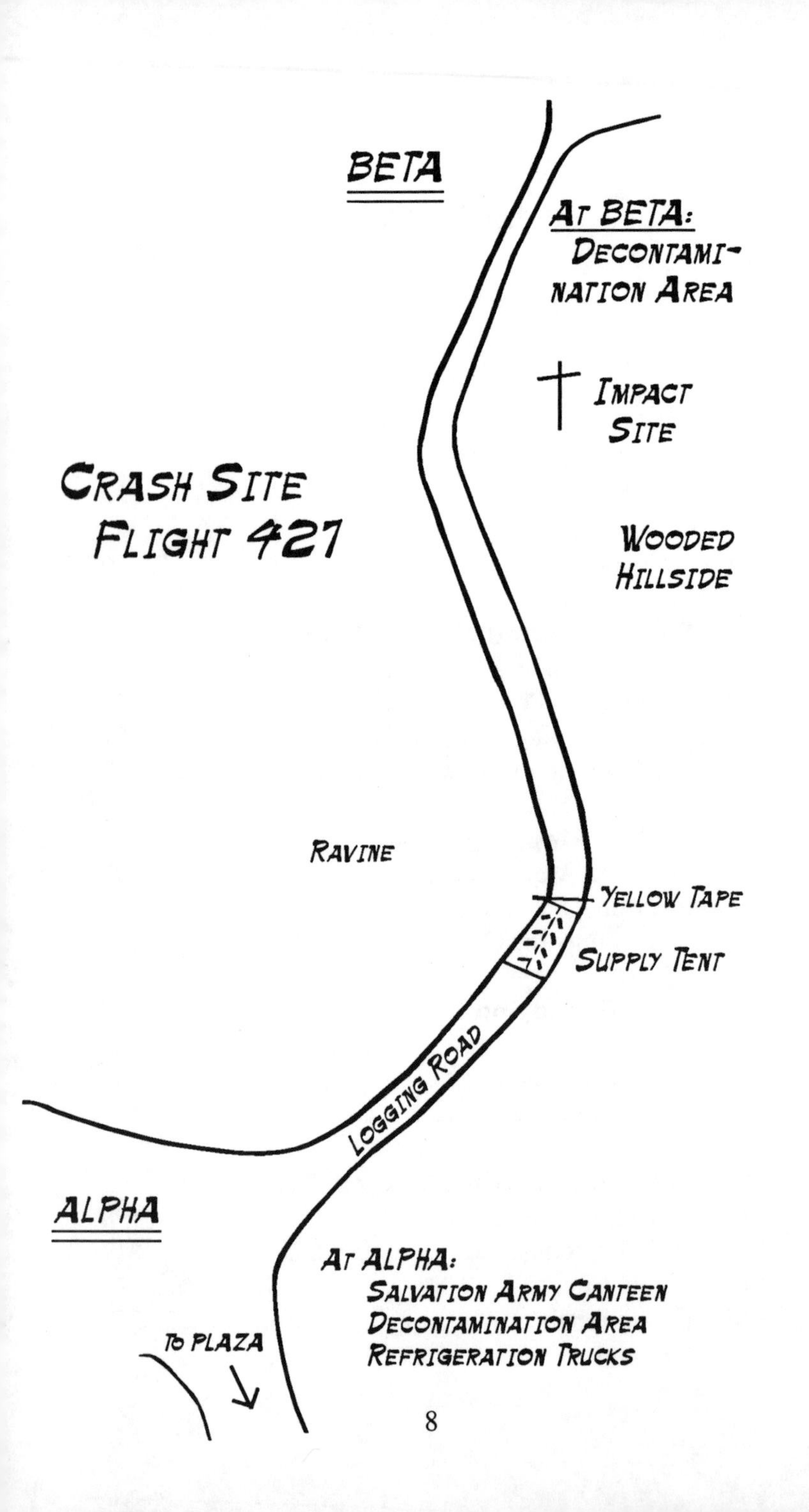
BETA
AT BETA:
DECONTAMI-
NATION AREA
IMPACT
SITE
CRASH SITE
FLIGHT 427
WOODED
HILLSIDE
RAVINE
YELLOW TAPE
SUPPLY TENT
LOGGING ROAD
ALPHA
AT ALPHA:
SALVATION ARMY CANTEEN
DECONTAMINATION AREA
REFRIGERATION TRUCKS
TO PLAZA

PROLOGUE

"Tell Me, Why Did This Happen?"

My hands played with the yellow tape.

The crash site was closed. We were not allowed to cross beyond the tape. Behind me, a team of military personnel was marching out; its task was to secure the perimeter of the crash site. A man from the Salvation Army pointed up the old logging road directly in front of us and said, "You can't see the impact site from here. But do you see, as the road bends to the left just up the hill, the burned trees? Right under those trees. That's where the plane crashed."

I tried to lift my eyes and follow his directions, but I

couldn't. The land was riddled with debris, as if boxes of dynamite had exploded, leaving nothing recognizable. I tried to look generally—not wanting to see the unbearable. But it was not possible. There in front of me, only ten yards away, was a mound of human flesh—and I couldn't take my eyes off it. The form of the man was gone. His hand was intact, but separated, entirely on its own. The brown pants and black belt stood out. The rest of the remains were undefinable from this distance. In my experience, I could only compare the charred body to the flesh of animals seen on a highway—utterly mutilated. But the hand was like my own.

That's what got me, and it tore at my soul. One long look and I saw myself on the ground, piled in a heap. This unimaginable human suffering—beyond anything I had ever seen or known—became intensely personal. These remains were human like me. This man could have been me. I grieved his death.

We were not at the crash site long. The truck took us back to the Green Garden Plaza, the headquarters for the emergency rescue operation. It was already teaming with media from all over the world. One step out of the truck, and we were swarmed by a mass of reporters with cameras, lights, and questions.

A man approached me from a local television news station, his cameraman behind him. He whipped off a question and stuck his microphone in my face. I heard him say, "What did you see up there?"

I had never been on live television before. I had never witnessed the violence nor the brutal suffering of what was all too fresh in my mind. I had no idea how to put what I had just experienced into words. I fumbled for something to say. *What did I see?* I wondered, *Who is watching this program? Are the family members of the 132 victims glued to their sets, hoping beyond hope? Did this man want physical descriptions? But how could I give that?* I felt the temptation to sensationalize and refused it.

Prologue

"The Beaver County coroner has closed the crash site," I started. "Yellow tape is blocking the entrance until the National Guard has determined the exact boundaries of the crash. We did not go in. We stood at a distance. But we saw enough, and let me say this: No one could have survived the impact. Debris from the plane is scattered everywhere. The destruction of Flight 427 is beyond description."

The reporter turned to my assistant, the Rev. Ken Ross. He wanted more raw description and possibly more human emotion. Ken held the line, giving only generalities and not succumbing to offering graphic detail. Then the man turned to the Rev. John Rucyahana, our African missionary friend and Anglican priest, who had spent that summer of '94 in his homeland of Rwanda. His people had just experienced the horrors of civil war and, worse, genocide. An estimated one million people had been hacked to death in two months' time. He had seen mutilation before. John begged the television viewers to pray to the Lord for the victims' families and the rescue workers.

Then the reporter came back to me. Ours was one of the first teams to go to the site. Eyewitness reports were still making the headlines this soon after the crash. But more, he had a new angle. We were not firemen, county officials, coroners, police officers, or National Guard members. From any of these sources, he could get a description of the wreckage. But we were different; we were clergy. He had one question, the inevitable question ministers are expected to answer in times of crisis.

"Tell me, why did this happen?"

At 7:03 P.M. on Thursday, September 8, 1994, a Boeing 737-300 from Chicago was on approach to Pittsburgh. There wasn't a cloud in the sky. On a soccer field directly below, the Hopewell Township Area Soccer League was getting ready for the start of their season, which would begin that Saturday. The field was flooded with coaches, parents, and nearly two hundred boys and girls, ranging from ages four to twelve.

As the plane passed overhead, some 5,600 feet up in the late summer sky, no one sensed danger. On the ground, they practiced for their match. In the air, the first officer told the passengers to prepare for a routine landing. He instructed the flight attendants to get ready for the plane's final approach to Pittsburgh International Airport. And then, just as the aircraft passed over the soccer field, a sudden jolt hit to the left, followed by another. Within twenty-three seconds, the plane bolted to the earth. The people on the soccer field didn't know anything had happened until they heard a terrifying explosion from the hill behind them. And all they saw was dark black smoke.

The next morning, just after ten o'clock, I was being interviewed on live television to explain why this plane fell from the sky. I'm not a plane mechanic, nor am I a Boeing engineer or a pilot. I don't know the technical reasons why Flight 427 crashed. I am a clergyman. The question posed to me was nothing less than a question about God Himself. The reporter should have asked it more clearly: "Why did God do this?" or "Why did He let it happen?" That's what he meant, and I knew it.

At once, I was being asked to make sense of this chaos. The question, phrased in this way, is unanswerable. It assumes that God is in control, holding complete authority over all things. In His hand was the power to prevent this tragedy, which He didn't do—*why not?* Didn't He love those people? Or maybe He wanted to save the plane and its passengers but couldn't. If so, then He's not all powerful, and that, too, is wrong. If God is God, then He ultimately must be responsible for the 132 people who met a quick, horrible death. At least that's how it appears.

The typical response of a clergyman in such situations usually satisfies the reporters, but really says nothing: "We don't always know why tragedies happen in life. We must accept them as one of life's great mysteries."

"Tell me, why did this happen?"

With the microphone pressed close to my lips, I looked the reporter square in the eyes and answered the real question: Where is God in human tragedy? I had only a few seconds, and I had to be clear. See, I am first a Christian, then a clergyman. And Christians should know how to answer that question. To me, it is the single, dominant, pressing theme of the Bible. It's the only reason Good Friday took place. There is an answer to this question, and I find it vital that those who follow Jesus Christ as their Savior and Lord know it and be prepared to offer it. The world is looking on. They want to know why. We must be ready.

Inevitably, the Crisis Comes

Let me ask a different question: What is the meaning of this short life?

Our busy, fast-paced American culture has no time for such a question. Our schedules are jam-packed. We don't even have time for ourselves. We're pulled in every direction just to make ends meet. Today, both parents work full-time jobs, and children are left to grow up without their constant presence. For some of us, it's worse. We're faced with the burden of being separated and single. We are left with the mounting guilt of not being present for our kids, stress at work, financial demands, social pressure from friends, and the inner demands—challenging and never ending—to be successful, attractive, pulled together, and in control.

When we have time to ourselves, what do we do with it? It's far easier to fill our silent moments watching the blaring TV, talking on the phone, or falling tired and helpless into our beds. We don't take time to think, dream, read the Bible, pray, and wonder about the meaning of life—especially *our* life.

Nothing is simple anymore. Our world is complicated, and even relationships are difficult. How do we love someone else when *we're* so mixed up? We don't know who we are or what our purpose is on planet Earth. Love and being loved have become illusive concepts, built on feelings which soar with satisfaction and ecstasy one minute, and then crash headlong into uncertainty, confusion, and despair. We fall out of love when our needs aren't met, and the vow we said before God, "Until death do us part," means nothing.

We want life to be manageable. We want to be happy, have the right person at our side, find the needed time to be with our children, enjoy enough money to live as we choose, be satisfied in our work, go to the perfect vacation spots, and even take an early retirement. Life, we think, is meant to be positive—full of possibilities. We feel we deserve only the best experiences which, from time to time, get tripped up by unfortunate negatives. But overall, we expect to grow up, become successful, make a name, and live a full, healthy life into old age. We want this for ourselves and for our children—and nothing less.

When tragedy strikes, we're surprised. We don't understand it when a child dies in our arms, a fire burns down our house, or a boss demotes us—or when we end up hooked on alcohol and drugs, or find ourselves divorced, alone, and out of control. When life turns sour, we say, "That's not what life was supposed to be!" What happened to positive thinking, positive living? Life is meant to be upbeat, hopeful, and altogether happy. It's the joy of wedding days, newborn babies, first days at school, graduations, promotions at work, building a retirement home, taking a vacation to Europe, buying a new car, and being together at Christmas with those we love. Life is a like a greeting card that wishes us a peaceful today and a promising tomorrow. It's not about sickness, crisis, or death.

When anything negative happens, we're shocked. We don't expect to face terrible tragedy.

Why?

I grew up in a family that worked hard at making the best of life. We were meant to grow up and embrace the challenges—to run the race as best we could, and to help others along the way. Sure, there were setbacks. But they were only stepping stones to make us stronger and of greater character. In the end, I was told, all of life's problems would be surmountable.

When trials did come, we as children were usually protected from the news. If a relative or dear friend took seriously ill, we weren't told until the person recovered. My family loved me and wanted me to feel safe. "We didn't want you to worry," they would later say. My parents did their best to protect me from the evils of this world and the terrifying lightnings and thunders of life. And I liked that.

But they couldn't shield me completely. My grandmother Jesse died when I was six. When I was eight, my grandfather, a man of great stature and distinction, suffered a disabling stroke. He lived eight more years, crippled in body, yet fighting to regain his quickness of mind and wit. But it was hard to do. Smaller strokes hit, one by one, and each time he faded from us a little more. When I was twelve, my sister, Kate, had major surgery. Two years later, my mother and father were separated. At sixteen, I stood in front of my mother's casket. She had died of cancer. As much as my family wanted to protect me and teach me that this life is full of adventure and pleasure, the darkness and thunder came—hard and strong. I wasn't prepared.

Dr. Martyn Lloyd-Jones, perhaps one of the greatest preachers of this century, was a trained physician who left his practice to take up the pulpit. Once in a sermon he asked his listeners to imagine a man suffering from acute abdominal pain. What would happen, he asked, if a doctor, concerned for the man's immediate condition, treated the problem by injecting the patient with morphine? This act alone would only relieve the symptoms. But it would not deal with the actual cause. For example, if the man had acute appendicitis, the morphine would disguise the seriousness of the man's state. He'd leave the doctor feeling well, but in reality, his appendix would soon burst. The result would be catastrophic.

"It is criminal," Lloyd-Jones said, "to remove a symptom without discovering the cause of the symptom." With that, he applied his analogy to a diagnosis of our culture's most dangerous disease:

> This "affluent society" in which we are living is drugging people and making them feel that all is well with them. They have better wages, better houses, better cars, every gadget desirable in the home; life is satisfactory and all seems to be well; and because of that people have ceased to think and to face the real problems.[1]

The reality is this: life isn't always rosy. It isn't a bowl of cherries or a smiley face wishing us a brighter tomorrow filled with sunshine and rainbows. Life is full of tragedy, pain, hatred, sin, evil, jealousy, sickness, natural disasters, and death. What is worse, we don't know when any of these might strike. There are no promises, no guarantees. One minute, all may be well—and then, twenty-three seconds later, we're face to face with our own death. On our very best days, there is always the possibility that the phone might ring with bad news, turning our lives upside down and changing us forever. As a pastor, I'm used to these phone calls: "The doctors said it was a massive heart attack...There's been a terrible car wreck...The lab tests came back positive...My boss just fired me...My wife's gone...We lost the baby...I heard gunshots. I'm scared."

One minute, our world is full of hope. The next minute, our heads are spinning. It's too easy to live in a morphine-injected state, seeing life as children might see it. Wrapped in our safe cocoons, we make believe that such phone calls will never come to us, that we are somehow protected from what other people suffer. We hurry through life believing (and teaching our children) that "life is satisfactory and all seems to be well." Consequently, we stop thinking, stop asking important questions, and utterly avoid the real nature of this world, which is full of human sorrow and tragedy. As Lloyd-Jones said, we never "face the real problems."

But one day, our safe, happy, pretend world will most certainly collapse. No one likes to say this, and it's hard to hear. But this earthly journey is all too real, all too filled with suffering. We can hide our eyes for only so long. We can think positive thoughts, whistle a happy tune, and hope against hope that our futures are nothing less than bright. But that simply isn't realistic. It's like morphine—a numbing drug that disguises the reality of the disease. No matter how hard we try or how desperately we shield ourselves and our young, it will happen.

Inevitably, the crisis will come.

A Response to God

My perfect cocoon, nestled in the bosom of a loving family that shielded me from dangers and things that go bump in the night, ripped open when I turned fourteen. Divorce had come to many of my friends' parents. I never once thought it would come to us, and yet it did. Dad moved out of the house. A short time later, he took a new position in San Francisco, which seemed like light years from our home in Pittsburgh. My brother, Gregor, was nineteen and went to live on his own. We sold our suburban house, and Mom, my sister Kate, and I moved closer to the city. Nothing felt safe or protected.

On my sixteenth birthday, we buried my grandfather. The eight-year battle of paralyzing strokes was over. He was a brilliant man, an entrepreneur who turned ashes into gold, and a man who dearly loved his family. Even at the end, with slurred speech and an old, useless body, his tear-filled eyes and strong left hand expressed the deep feelings of love he felt for his wife, children, and grandchildren. His death was the most painful death I had experienced until then. I hated it, and I missed him.

One month later, Mom went into the hospital for a routine gallbladder operation. The surgeon opened her up, saw her stomach full of cancer stemming from the liver, and sewed her back up. There was no way to remove it. He gave her six months. I didn't find out until mid-August, more than a month after the surgery. I had spent my sixteenth summer at a boys' camp in northern Vermont.

I had never known a better summer, filled with friends, laughter, and the high competition of tennis, sailing, water sports, golf, and softball. I left that paradise and took a bus to Boston to meet my Uncle Hays—an Episcopal priest and the dean of Bexley Hall Seminary in Rochester, New York. It was his job, on our drive from Boston to Newport, Rhode Island, to tell me that my mother, his sister, was dying.

Once in Newport, I went down to the beach by myself. It was a late afternoon on an overcast day, and the ocean wind was blowing strong in my face. The breakers on Newport's well-known "Second Beach" were crashing down in salty, white foam, hundreds of yards from shore. I walked until I found a place where no one could hear me or see what I was doing. And there, I looked up into the sky and prayed to God.

A devout Episcopalian, I had gone to church all my life. At seven, with love and admiration for my Uncle Hays, I had told my father and mother that I, too, was going to be a priest. Not once, growing up, did I lose the spark for that dream. But that afternoon on the beach, I didn't understand anything. How was it possible that my mother was dying? Why was this happening? How could her cancer be inoperable? There had to be a drug, a new medicine. Maybe there was a mistake, a misdiagnosis. Maybe God would hear my prayers and heal her. This couldn't be happening. No, Mom wouldn't die. I wouldn't let her.

I stood on the beach, tears running down my face, and cried to the Lord with a loud voice, "Help me!" Mom needed my prayers. But when the words came, all I could think about was myself. My world was collapsing and I could feel it. This news was bigger, greater, higher than my heart and mind could comprehend. "I don't get it!" I shouted. "How can Your world include cancer that takes my mother away? Is this Your plan for her life and mine? Who are You to let this happen?" I was lost. My soul, breaking under the weight of this news, begged Him, "Please, don't take my mother. Do something. Don't let her die."

There on the beach, I turned to the Lord. I didn't know Him very well. I had never read the Bible, nor did I understand His purposes and plans in life. In fact, I wasn't even sure what it meant to be a Christian. I just went to church and wanted to be a priest. But deep inside, I felt that I could trust Him. He was God, "Maker of heaven and earth." All things were in His control. He could hear my prayers. He knew my heart and He knew my mother. And I had no one else to turn to—no one like Him, and not in a time of crisis like that.

The Other Response to God

Not everyone turns to the Lord in crisis. Some turn against Him.

Jacqueline was my first best friend. She lived a few doors away and across the street. Our parents were close; her dad is my godfather. She and I often played together until my family moved away when I was six. But our friendship continued over the years. Her parents would come to see us, or we'd return to our hometown. As far back as I can remember, I knew that Jacqueline was a miracle child. She was not easily conceived. And when she finally came, she suffered from severe allergies, which, on more than one occasion, threatened her life. Growing up was not easy. But as she entered her teen years, the allergies slowly disappeared, and Jacqueline blossomed into a healthy, beautiful, gifted young woman.

I will never forget her tears at my mother's funeral. There is nothing quite like the comfort of your oldest friend bearing your same pain. She loved my mother and told me she couldn't understand why Mom had died at forty-five. It was too young, she said. It wasn't right. It wasn't fair.

Jacqueline was not yet twenty when she went out west. She had finished her first year at Colgate University and had the opportunity to take a summer job in Jackson Hole, Wyoming. The adventure of leaving home, taking a job in a remote place, and being with friends for the summer excited her. It was all part of growing up. But that summer met with a tragic end. One night, after a party, Jacqueline and three other friends were driving home. She was in the front passenger's seat. The driver had drunk too much and his judgment was impaired. The car swerved off the road and crashed. Hours later, the phone rang at her parents' home outside Detroit. The driver and two other passengers survived. Jacqueline was dead.

Her mom and dad asked me to be a pallbearer at her funeral. Her death didn't seem real to me. I kept looking at her parents. I couldn't imagine their grief. What was it like to get that phone call, to meet the plane that carried her dead body, or worse, to kiss her good-bye? How could they live without her?

At Jacqueline's funeral, one of her classmates took me aside. She was the first person ever to confront me with the question of why: "You're a Christian, aren't you?" I told her I was. "Jacqueline told me about you. You want to go to seminary and become an Episcopal priest, right?"

"That's what I want to do."

"So then, you explain it to me. Why did this happen? What kind of God would let a nineteen-year-old die? Isn't He supposed to be loving? Then what? He didn't love Jacqueline? She was the best person in the whole world. So why did He let her die? We were best friends. She had her entire life in front of her. And what happens—the other three walked away, including the driver. I think it's completely unfair. I want her here. She was your oldest friend! How can you go and give your life to God now, after this?"

I had nothing to say. It did feel unfair. Her death hurt so much.

Jacqueline's friend was the first. Over the years I've met many people who blamed God for their tragedies. They ask why, but down deep they can't hear the answer. In their minds, God did it. He caused the death of someone they loved, leaving their hearts broken and empty. Who else is there to blame? In bitterness of soul, they turn away from the Lord, saying, "God, why have You done this?" In time, if no proper answer comes, they remain separated from God, left to carry the heavy chains of loss by themselves.

There are only two responses in times of tragedy. The first is like that of a man in the Old Testament named Job. When death came to his children and sickness to his body, he turned to the Lord, saying:

> Naked I came from my mother's womb,
> And naked I shall return there.
> The Lord gave and the Lord has taken away.
> Blessed be the name of the Lord (Job 1:21).

The second is like Job's wife. She turned away and blamed God in her loss, saying to her husband, "Curse God and die!" (Job 2:9). These are the people who want to put the Almighty God on trial for charges that He's responsible for their loss. He must answer or else be found guilty and held accountable in their eyes. Such people become the prosecuting attorney, judge, and jury. Their pain is blinding. They need answers, and someone must be blamed. God is the logical choice. In their view, He is a monster, causing pain at will and striking down in death with no rhyme or reason. He is unjust, unloving, and downright mean.

But that's not true. There *is* a reason why. It's just hard to hear when the heart is crying so loudly.

There's a Time to Hear Why

At funerals, especially in cases of early or tragic death, I am often asked the question Why?

Some people I call *outsiders.* They aren't directly affected by the death, so their grief is on the surface. They might pop the question at the funeral home. I always answer them, but I can usually tell whether the question was asked for academic, philosophical reasons or because their souls yearned for heavenly truth. Jesus said, "Ask, and it shall be given to you; seek, and you shall find; knock, and it shall be opened to you" (Matt. 7:7). Outsiders don't really seek an answer. They're looking for the thrill of the discussion itself.

Insiders are those suffering directly from the loss. Often they don't want an answer either. The shock is too real, the pain too hard, the grief too deep. Nothing will satisfy them but one thing—to have their dead back. That's why comforting insiders is so difficult. We say, "At least the suffering is over." Or, in another situation, "We can be thankful it was a quick death." But words just don't work. No matter how old the dead person was, no matter how sick or how long the suffering, those who truly loved them just want them back—alive, healthy, forever. Death has come like a thief in front of their very own eyes. Their hearts are torn in pieces, and their normal, everyday lives are left empty and abandoned.

No answer satisfies. Not then. Not for insiders.

The question of why is best asked before the crisis comes. But we don't tend to do that. As a pastor, I see most of us managing our lives in a secure, regular routine. Sometimes God is part of that routine. We say a morning prayer in the shower and go to church on Sunday mornings. But even then, church can be a low priority on our week's "to do" list. We're so busy with family duties, friends, out-of-town visitors, Saturday night parties, yard work, and of course, "Sunday is the only morning for us to sleep in." For some of us, He's there, part of life's hectic routine, but relegated to unimaginable unimportance.

Until the crisis.

It is common to avoid staring in the face of death until we've got no other choice. When I was growing up, we weren't allowed to see the dead—only closed caskets. I was in my twenties when I first saw a dead body. Others grow up seeing the dead in open caskets with their faces made up, hair styled, glasses on, wearing nice clothes from home, and with their hands neatly folded. It appears as if they're sleeping. I am always at a loss when someone says to me, "Oh, doesn't she look nice!" I smile back, but inside I want to say with sarcasm, "No, she doesn't. Not at all. I found her looking much nicer when she was alive and breathing." Right?

Denial is present in both cases. Death is hard to face. It reminds us, like a thorn in the flesh, that we ourselves are going to die. Who wants to think about that? But how can we truly live until we have come to terms with our inevitable death? I ask people, "Are you ready to die? Are you ready to meet the Lord, the Judge of all the earth, and give an account of your life?" I find most people unable to answer me. They don't want to talk about suffering and death, especially their own. It's far easier to pump morphine into their bodies and imagine life full of fresh roses today, chocolate-covered tomorrows, and no death.

But there comes a time when we must face it. We must ask, "Lord, where are You in tragedy?" It's best to ask this before we face our great tragedies. If we don't, the next time to ask—and I mean with a burning fire in our bellies and a desire to hear His answer—is in the middle of the grieving process, sometime after the shock has worn off.

There is a time, a right time, to hear what the Lord has to say. And He does have something to say.

His answer is nothing less than His story. The question Why? is the single question Jesus Christ came to answer. The Son of God descended from His eternal throne of glory and was born in Bethlehem, in a manger, of a human mother, just to answer our heart's deepest cry in this suffering, death-stricken world. We may ask the question and not hear His answer. But He does not avoid the question. And He never has.

No More Morphine!

When I stepped onto the crash site that first time, it was as if I had never tasted the death of another before. The anguish was far worse, far more painful than I had ever known. A thousand times, I wanted to run from the site, bury my head, take the morphine into my veins, and pretend these deaths never happened.

The only reason I stayed was because I knew His story.

I knew that the Lord never runs from tragedy. He doesn't stand on the outside looking in. He knows our suffering. He promises to be with us. He is able to face our accusations, which hold Him accountable for our loss. No matter how deep the pit, how dark the circumstance, how strong the storm, the Lord has sworn that He will not leave us. There, even in our darkest tragedy, when we are reeling in grief, lost and out of control, He's there for those who call on Him. And when we're ready, He is most willing to tell us why.

---

[1]D. Martyn Lloyd-Jones, *Preaching and Preachers* (Grand Rapids, Mich.: Zondervan Publishing House, 1971), 31-32.

Part I

Seeking a Safe Place

Chapter One

Thursday Night, September 8

Anna had slipped into a coma. Her daughters, Bev and Betty, her son, Georgie, and their families stood around her bed. At times, the hospital room was filled with delight as one of them told a particular story, detail by detail, revealing Anna's maternal strength and humor. She was, even in her nineties, the clear thinking, driving force of the family. Everyone knew it, and everyone knew she was dying. With that thought came the tears, then the silence of unbelief, embraces of comfort between them, and a determined resolve not to let their mother and grandmother go. Not just yet.

But their own resolve betrayed them. It was time to let her go. Erilynne and I watched as Bev and Betty stroked their mother's hair. Georgie held her hand. Each one spoke his or her last words to their mom. Anna was quiet and peaceful most of the afternoon, her condition steady. Near five o'clock on Saturday, September 3, I heard Bev cry out, "Mom!" Her breathing had slowed. A breath, then a pause. A breath, then a longer pause. Anna's time had come. Spouses and grandchildren huddled close around the bed, watching and waiting, until a few minutes later when there were no more breaths.

Anna's family would never be the same. She was the centerpiece, like a precious jewel that adorned their lives with richness. Their Sunday night family dinners would go on—they pledged that to each other. But how empty they would be without Anna! The photograph by her hospital bed, taken two months before, showed her happy and well, sitting in a summer chair outdoors and holding a newborn great-grandchild in her arms. But Anna now lay quiet, that fervent devotion to her family gone like that hot summer day.

For a brief moment, Erilynne and I felt as if we were part of Anna's old-fashioned family. In this modern age, it's customary to find children living far from home, with parents separated and older people gently pushed aside from the rushing tides of family activities and schedules. But here was a family who stayed together and loved each other, and they now fell weeping into each others' arms. What an extraordinary gift for Anna, I thought, to die in the warm, soft embrace of her loving family. Not alone. Not pushed aside.

The funeral was held on Wednesday, September 7. Wayne Tatalovich, his daughter, Teri, and his son, Wayne, were the funeral directors. It had been a year since we had led a funeral service at Tatalovich's. I remember Wayne Sr. being frustrated that day. He's a handsome, slender man in his late fifties, with an unbridled ambition for the golf course. Somehow he had injured his arm. There he stood, with one arm bound in a sling and his face slightly downcast from his doctor's orders: No golf for a few weeks.

Anna's funeral was a sad one. No one had wanted this

blessed woman, even after ninety-plus years, to leave them. When it was all over and the family had gone, Erilynne and I shook Wayne's good hand and thanked him for his services. We chatted casually, then he got into his limousine and drove off—Wayne Tatalovich, funeral director and elected Beaver County coroner. Little did we know then that we'd see each other the next day, that awful day, and that our lives would never be the same.

At Music Rehearsal

Thursday, September 8, was a dreary day. As I drove to the church in the late afternoon, my mind went back over the week. Not only had Anna died, but two colleagues in ministry had also died. I'd spent the morning in Pittsburgh at the funeral of one. He was in his early fifties—a massive stroke. The other man was a good friend, the Reverend Mike Henning. He excelled as a preacher. His last sermons had moved me deeply as he pleaded for people to turn to Jesus Christ while there's still time. He died of cancer at the age of forty-five.

I pulled into the church parking lot at 6:15 for music rehearsal. The church my wife and I pastored, Prince of Peace Episcopal Church, was a young, exciting congregation. It was only nine years old, having started in the Hopewell Township fire hall in 1985 and then growing through two building processes. Our church family had a zeal for the Lord, to worship Him and to hear His Word. Our music team was composed of an organist, guitarists, a bass player, and a number of vocalists. We usually prayed together at 6:30, talked for a while, and began our practice for Sunday's worship.

I looked out the window during one of the hymns. The night was postcard perfect. Not a cloud in the late summer sky, nor a breeze rustling through the willow tree some one hundred yards away. My heart weighed heavy as the beauty of dusk fell over us. It had been a sad, tiring week.

In between our songs of praise, just past seven o'clock, we

heard the local fire station, a block away, sounding its siren. We kept right on playing—it was a common sound to us. Soon we could hear sirens blasting in the distance and fire trucks blowing their horns on the main road in front of the church. With the emergency sounds outside and the music practice inside, we couldn't hear the phone ringing in the kitchen at the back of the sanctuary. We later learned that family members were trying to call us.

On we sang. At one point, Mary Jo Borchart, our pianist, said, "It must be a terrible fire." We had never heard so many different kinds of alarms wailing and echoing all around us. Was the fire nearby? Was it the home of someone we knew? Should we go out and see? What direction were the trucks going? For whatever reason, we decided to keep playing even as the sirens blared louder and louder.

At 7:15, Johnene Belsky, a USAir flight attendant, came running into the sanctuary and up the middle aisle. Her husband, Jeff, stopped strumming his guitar. He could see the panic in her motions, and as she came closer, the redness in her eyes and the tears running down her cheeks. Immediately, he thought of their two-year-old daughter, Jordan. Had something happened to her? The music stopped by the time Johnene got to the front row and collapsed into one of the chairs.

By then, we all knew something was wrong.

"A jet went down."

The words came but not the understanding. My mind couldn't register that the sounds of sirens shrieking outside and her simple, tearful words were somehow linked. Jim Rosenkranz, the bassist and a DC-9 pilot for USAir, got it immediately. He knew the feared words. They were part of his life's work. He put his instrument down and asked, "Was it a USAir jet?" His mind raced to all the pilots and crew members he had flown with over the years and knew so well. Were his friends on board? Were they suffering?

The questions started coming at Johnene from all

directions: Was anybody killed? Did people die on the ground? How did it happen? Was it taking off? Was it landing? She couldn't answer. All she knew was that the plane had crashed near our local shopping center, the Green Garden Plaza, about a mile from the church. The first reports gave no hope for survivors, but they were early reports and all too sketchy.

I turned to look out the window. The sound of the alarms still filled my ears with a terrifying violence. They were saying the same thing: A jet went down. Over a mile away. Crashed at the Green Garden Plaza. Rescue workers were heading to the scene. It was all too real, all too close.

The Black Smoke

Practice was over.

Jeff went to his wife and comforted her. I looked at Ken Ross, the assistant pastor, and said, "We've got to go to the Plaza." Jim said he wanted to go. Jeff offered to drive. Christine McIlvain, one of the singers, said she'd close the church. The rest of the singers and instrumentalists sent us out, promising their prayers.

The Beaver Valley Expressway was the fastest route. The Plaza was right off the Aliquippa exit. But as we came near the expressway, we saw policemen blocking the entrance. Jeff quickly turned right onto Gringo, a winding country road that eventually comes out at the Plaza. The road was open. We sped down it, still stunned with the news. USAir did not need this crash. Only two months before, a DC-9 had gone down in a thunderstorm at the Charlotte, North Carolina, airport due, at first reports, to a wind shear. How could this airplane have gone down? It wasn't weather—the night was perfect. We turned to Jim. He had the flying experience—what did he think? "Right now, it's anyone's guess," he said. "The flight data recorder should tell us something. We'll have that information in a couple of days."

It didn't make sense. Not to me.

Jeff had a car phone. We decided to call Bob Dwyer, a dear

friend and a thirty-year employee with USAir. He worked in systems control, and I was confident he'd know the story. The phone lines were jammed. I kept pressing redial and got him on the fourth try.

"Right now, we understand this was an inbound flight from Chicago to Pittsburgh," he stated with confidence, trusting his sources. "It was on its final approach when it disappeared from radar. I haven't seen a report from the tower yet. I don't know if they were aware of any problems on board the aircraft." I could hear the shock in Bob's voice—and the strength. I knew the weight of responsibility facing him.

"What else do you know?"

"This was a Boeing 737-300 with 126 passengers and six crew members. Also, the crew was not from Pittsburgh. They were out of Philadelphia. I am sure of that. Also, the plane was right on schedule."

"Bob, are there survivors?"

"We don't know."

"And does anyone know why this plane went down?"

"Not that I know of—not yet. Lots of rumors, but don't believe them. It's too early to tell."

When I hung up the phone, we caught our first glimpse of the Green Garden Plaza in the distance. In my mind, I kept seeing the downed plane in the actual Plaza parking lot. Shoppers would still be at the Giant Eagle grocery store, the Revco pharmacy, the New York Pizza and Pasta restaurant, and the other various shops and businesses. I kept wondering how many were killed on the ground. But as we looked, all we could see were countless flashing lights from the emergency rescue vehicles. There was no sign of the jet.

As we drove closer, Jeff was the first to see the smoke. Behind the hill, rising up to the right of the Plaza, was a dark black column of smoke, gushing upward into the deep blue twilight sky. As we arrived at the entrance of the Plaza, cars were being held back by local policemen. We got in line. To our right, across from the Plaza, dozens of ambulances from all over the Pittsburgh area were lining up, ready for action. Their flashing strobe lights hurt my eyes. The ambulance workers were huddled close together, waiting.

It was our turn to face the policeman. Jeff rolled down his window.

"This is for rescue operations personnel only," he said. I leaned toward Jeff and caught the man's eye. "Clergy," I stated factually. He saw my black shirt and white collar. With nothing more said, we were waved into the Plaza. We turned in and saw a vast array of emergency vehicles and rescue equipment of all shapes and sizes. Jeff made a left and found a place to park. Ken and I got out, leaving Jeff and Jim at the car. Off we went, but where were clergy needed?

Instinctively, I thought our best position would be with the medical teams. They would be dealing with the survivors, the dead, and possibly family members. It was hard to know where to go. Fire trucks from over forty fire departments and police cars were still flooding into the parking lot from every neighboring county. It was a frenzied chaos of sirens wailing, lights flashing, helicopters flying overhead, cars and people circling everywhere—and that deadly smoke.

I could not take my eyes off that dark black smoke.

We found the medical personnel. They were grouped in one section of the parking lot. Like everyone else, they were waiting for orders. But from whom? Who was in charge? We were told that the Unis car dealership was being readied to receive the family members of Flight 427 passengers. A temporary morgue was being set up a few hundred yards beyond Unis. Local stores were donating food and supplies that might help the rescue effort. As night fell on the Plaza, I was awestruck to see the immediate response of the men and women of Beaver County and her neighboring communities. These brave people were there when disaster hit, fully ready to go to the crash site, serve the wounded, and care for the dead.

"My husband was on that plane!" I turned around and saw a dark-haired woman coming toward us in hysterics. "I took him to the airport an hour ago. He was headed to San Francisco. On the way home, I stopped at the grocery store. Then I heard the explosion. Oh, my God, this was his plane!"

"I heard it was going to Los Angeles," one man told her.

"Yeah, I think eyewitnesses said the plane was taking off," a nurse to my left added.

"I heard it was coming into Pittsburgh and had to abort its landing," another man reported.

"Actually, no," Ken came in with authority. "We talked with a man at USAir systems control only a few minutes ago. He confirmed that this was an inbound flight from Chicago with 132 people on board." Then he turned to the crying woman and said, "I can assure you, your husband wasn't on this plane."

In the course of the next hour, we talked with many such people. They came out of the shopping center fearing the worst. Our Hopewell Township community has many USAir employees. Pilots and flight attendants leave for two, three, or four days at a time, flying to many cities. Their family members often don't know where they are at any given moment. It was possible that their flights could have taken them back into Pittsburgh, a USAir hub, and then rerouted them to somewhere else. Easily, it could have been them.

"We are certain," we repeated, "that this was a Philadelphia crew." That comforted the Pittsburgh USAir families, who for a moment thought their loved ones were on board. But the sadness remained for all of us.

We kept waiting, standing helplessly without orders and with nowhere to go. The more time passed, the more impatient everyone became. Why the delay? A reporter came over with his cameraman following closely behind. The bright light of the camera hit one of the doctors in the face by surprise.

"Doctor, what do you expect to do when you get to the crash site?"

The question was inane. He answered sharply, almost rudely. He wanted nothing to do with the press. He was there to save somebody's life if he possibly could. The plane had gone down more than an hour ago. Why was he talking to a reporter and not working with the injured? What was holding things up?

"What's your name, Doctor?"

His eyes were still on the hillside where the black smoke, now in the dark of night, could not be seen. He snapped back, "Who cares about my name?" The reporter got the message and backed off.

The doctor, like everyone else, was on edge. All the power and might of Pittsburgh's finest rescue teams stood at the base of the hill, ready to give the passengers and crew of Flight 427 a fighting chance to live. All that was needed was the command to go. Someone in charge needed to snap his fingers, give the word, and begin the rescue operation. But that wasn't happening.

Time was passing. The plane had been down too long. There were still no orders.

Word From the Crash Site

Ken and I meandered between the medical teams and the Unis dealership. One of us needed to be at Unis in case family members arrived, or down at the temporary morgue if bodies started coming down by helicopter or ambulance. The other had to be ready to go to the crash site with the medical teams. But there wasn't action anywhere. Once, a helicopter landed at the morgue. As we walked closer, we saw men loading supplies on board. Then the helicopter took off again. There were no bodies.

Rumor had it that one child had been taken by emergency helicopter to a Pittsburgh hospital. Amid the long waiting, it was the only sign of hope to boost our morale—*maybe there were survivors!* But why the wait? We finally found the emergency command headquarters for the rescue effort. The local officials were securely in charge of all the emergency personnel, their vehicles, and their deployment to the site.

Already, the Federal Aviation Administration (FAA) had charge over the airspace. I later heard from Terry Thornton, an FAA official who attended our church, that the Pittsburgh International Airport flight control tower had restricted all aircraft from flying near the crash site. The next morning, a small plane and a helicopter disregarded those orders. They flew in, probably for a press release photo. Both pilots lost their license. But the National Transportation Safety Board (NTSB), now en route to the scene, would ultimately be in charge of the rescue operation. They were responsible for the government investigation of the crash. There was a definite line of authority, even that night amid the seeming confusion at command headquarters, and I knew it.

"So, what's holding up the medical teams from going to the crash site? Who gives that order?" I asked one of the county officials.

"Wayne."

"What's his role?"

"County coroner. He's up there now with his team of coroners. We're waiting for his report. He'll make the decision. Once we get word from him, the medical units will be the first to know." Then he slapped me on the back, gave a half smile, and said, "I'm glad you're around, Padre. We need the clergy here." He quickly turned away and went about his business.

Wayne Tatalovich. All the ambulances, fire trucks, emergency vehicles and medical teams were waiting for his report. He was there at the crash site, calling the shots. That's when I got it. I knew some people thought we were waiting because there were no roads to get to the site. Others believed that helicopter medical technicians were flying the survivors out and didn't need our assistance. Still others thought no one was in charge and that someone, somewhere, wasn't doing his job. It was hard not to complain—just standing, waiting, with adrenaline flowing. It was 9:15. The plane had crashed just over two hours ago. But I finally understood. Our waiting made sense. Someone was in charge. Our county coroner was at the site, and he wasn't calling for emergency assistance. Why? He didn't need it.

Around 9:30, a group of cars entered the Green Garden Plaza and drove toward Unis. Ken and I made our way there. We overheard someone say, "Here comes Wayne." By the time his car had come to a stop, a crowd had gathered. As he got out and made his way toward the dealership, the media kept pressing in around him, asking questions. County workers surrounded him and cleared a path for Wayne to walk through the side door of Unis. He said nothing. He wanted to make his first report to other county officials before talking to the press. Ken and I stood still, waiting at the side entrance. For whatever reason in all the commotion, Wayne saw us before going through the open door.

"Where are the living?" I mouthed the words. I knew he couldn't hear me. But I gestured with my hands half raised in the air as if puzzled and asking a question. He saw it and seemed to understand. Back came three words—clear, pained, and hopeless.

"Pray for me."

There were no survivors.

Closed Until Morning

Ten minutes later, word sifted through the crowd. The emergency vehicles were no longer needed, the medical teams could go home, and the volunteers still willing to serve should come back in the morning. The crash site was officially closed. No one would be allowed up there until daybreak, save a group of Hopewell policemen who were assigned to guard the area from looters. They had the right hunch. Two looters, looking to steal from the dead, were caught and arrested in the early morning hours.

There was nothing more anyone could do. It was enough to know that 132 people (and two more, since two of the women were pregnant) were officially dead. The clean-up effort must wait for daylight. But more, a strategic plan of action needed to be outlined in detail by the county coroner and the Federal Aviation Administration under the direction of the National Transportation Safety Board. The work had to be done by the book, with hundreds of volunteers beginning at sunrise.

Somehow tragedy is more tolerable when lives are saved. The adrenaline rushes, the body is ready for action, the mind can't think fast enough, and the heart says, "I must do something!" But there was nothing to do. We had to go to our homes, sit in our soft chairs, and face the memories of the night. And for me, no memory compared to that of the black smoke billowing high into the air. It spoke of the deaths of 132 people who were trying to make their way from Chicago to Pittsburgh. It didn't matter whether or not I knew someone on board. That night, death hurt. Its sting drove deep inside, as if I had lost someone close to me.

Before Ken and I left the scene of the crash, I found a piece of paper to write on and walked to Wayne's car. I can't remember my exact words, but I wanted him and his wife, Valerie, to know we were praying for them. I wanted them to have confidence that the Lord would be with them. He never abandons His own, but is always present in times of tragedy and great responsibility. He would be their strength, their courage. I slid the paper into the crack where the driver's door opened, hoping he'd see it.

Jeff and Jim had waited for us. We drove back to the church on the winding country road—the expressway was still closed. There was little to say. We were all so somber. Ken and I had heard rumors that family members were still at the airport. We decided to go, thinking there might be something more we could do. Our trip was in vain. Family members waiting for Flight 427 had long since gone.

I drove Ken home.

"What about Sunday?" he asked. "What are you going to preach on now, in light of the crash?"

"I don't know. I hadn't thought of it."

"Prince of Peace is full of USAir workers. All of Hopewell

is shaken by this tragedy. I think you are going to have to deal with it somehow."

"It's going to be like a funeral, isn't it?" I said back.

We parted, making no plans for the morning. Neither of us knew what to expect. I got home at a quarter to eleven to find a group of people in our family room, glued to the television set watching the local news report on the crash. My wife Erilynne, John and Harriet Rucyahana (our missionaries from Africa), their teenage daughter Hope, Joyce Wingett (our friend and a staff worker at the church), and her fifteen-year-old daughter Donnica welcomed me home. They had stayed up to wait for me.

Erilynne, John, and Harriet had been at a dinner meeting. They had arrived home after nine and learned about Flight 427. "Mom, Hope, and I were eating dinner at the New York Pizza and Pasta restaurant at the Green Garden Plaza," Donnica said, her bright blue eyes wide open. "We were right there!"

"We heard the explosion," Joyce reported, "but we had no idea what it was. We ran outside with everybody else. Someone screamed out, 'A plane crashed! A plane crashed!' But we didn't believe him. I mean, come on! That's ridiculous. But then we saw the smoke coming up from the hill. Other people were saying the same thing. Within five minutes, the place was full of fire trucks, ambulances, and police cars everywhere. We decided we'd better grab our food and get out of the Plaza as fast as we could."

"We came right here," Hope said. "It was so scary. You can't imagine a plane crashing right outside your restaurant window. And that explosion was so loud. I still can't believe it."

"Did you hear about the soccer field full of children?" Erilynne asked. "The news is reporting that there were some two hundred children and parents at a nearby soccer field. This plane's flight path went directly overhead. They said if the jet had gone down three seconds earlier, the plane would have landed on the field, killing everybody. Now, that's even more terrifying to think about." She was right. It was the first good news of the night. The troubled jet did not land on the kids or at the Green Garden Plaza, but in an unpopulated, wooded area alongside an old logging road. No one on the ground was killed.

We prayed together before our company went home. There was so much to pray for that night. We remembered Wayne Tatalovich, his team of coroners, and all the policemen and firemen who had already been to the crash site and seen the destruction and death. We prayed for the families and friends of those on Flight 427 and for the USAir employees who, that very night, had to call them and tell them the sad news. We asked the Lord to be with all the rescue workers who had raced to the Plaza to give their time and help. We prayed for our church, the residents of our Hopewell community, and all who, in some way or another, were grieving from this horrible tragedy. This night was a sad night.

Planes have crashed before. Earthquakes, wars, murders, violence, sickness, and sudden and tragic deaths are common to our nightly news. We watch. We sleep. We're unchanged by disaster when it happens somewhere else. But when it comes near, it's different. We get undone. We feel the human suffering; we weep for the victims and their families as if they were our own. We lose sleep grieving the loss. We give our money, our time, our sweat. We get on our knees to pray, even if we've never done it before and don't know how. For some unknown reason, everything else is a just a television show—turned on and turned off, unreal and far away—until it arrives in our world and at our back door. And then it means something.

The crash of Flight 427 was not some television news story or a headline for tomorrow's paper. It was a real accident with real death, and it touched our lives. No, it didn't belong to the press. It belonged to us—the people who were there, carrying the sorrow and death of the 132 in our hearts.

Unsafe

I couldn't sleep. My mind raced through all the images at

the Plaza. I could still feel the loss of Anna and my friend Mike, who had died so young of cancer. As the darkness of night washed over me, the feelings of grief and emptiness grew. But they were confused and unfocused. Old griefs intermingled with the new, as if all of them were trying to roll into one at the same time and place, leaving me feeling alone and scared.

Drifting in and out of sleep, I found myself on board the plane—flying into Pittsburgh as I had done many times before. I was sitting on the right side of the aircraft in a window seat just behind the wing. I could feel the thrill of coming home, knowing that life, in general, was neatly tucked under my belt. I was healthy and blessed, with a beautiful, loving wife, an ideal job, a bright and promising future, and a safe, little world filled with pleasure and happiness. I took a quick glance out the window on a perfect, cloudless evening. I saw below me the familiar, rolling hills of western Pennsylvania drenched in a soft, golden sunlight that warmly invited me to come home. What more could I ask for?

The plane jolted to the left. Once. Twice. Then down it went, fast, like a roller coaster. No, that's not all. The screams, the terror. It's more. It's worse. Down, driving down, faster, harder. And then, all of a sudden, I sat up in my bed—caught between the nightmare and the reality, between my death and someone else's. My mind wrestled itself out of sleep. Yes, the plane did crash. No, I wasn't on it. I was home, safe. But those other people on the plane—the 132—were all dead, their bodies laying still on a hillside a mile and half from my home. That wasn't a dream. That could have been my plane.

Life isn't neatly tucked under my belt. It's fickle, frightening, and tragic. I lay back down on a cold, wet pillow that was soaked with my own sweat. The crash of Flight 427 had surfaced my worst fear.

I felt unsafe.

Chapter Two

GROWING UP PROTECTED

I was mowing the lawn when the beeper went off. Erilynne came and got me. I washed and dressed, got in the car and raced seven miles to the Stamford Hospital, located near our home in Connecticut. It was a late Saturday afternoon in July of 1985. As a seminary student, I was spending the summer as a chaplain-in-training. The beeper was part of the program. We were required to take it on alternate weekends, which meant we were on call for the emergency room. When an ambulance on the way to the hospital signaled "code red," the beeper went off. We were expected to drop everything, no matter what we were doing, and bolt to the emergency room pronto!

When I arrived at the hospital, a nurse told me that a forty-eight-year-old man was suffering from a cardiac arrest.

"He's in room one," she said, pointing over my left shoulder. "His wife and a neighbor are in the family waiting room. She's not in good shape. I think that's the place for you."

The door to room one was closed, so I quietly opened it and slipped in. I wanted some sense of the man's diagnosis before I met with his wife. Was he responding to medical intervention? Was his heart back in rhythm? Was their any optimism for the man's recovery?

I stayed by the door. I saw the man; his clothes were ripped off, and intravenous lines had been stuck into him with their bags swinging above. Monitors were blipping and beeping, and a mob of doctors and nurses moved around him in a frenzy. I heard the familiar word *arrhythmia,* which meant that his heart still hadn't caught its normal beat. Doctors called out orders with shouts, as if they were military commanders. A needle penetrated the man's skin, the injection made. Then another injection. Electric shock paddles came down on his chest and fired, making his huge body jolt from the table. Nothing. More orders. More aggressive measures.

I slipped out and headed down the corridor to the waiting room. It was an undecorated, small room—square, without windows, brightly lit, cold, and sterile. Two women huddled close together, each with both hands wrapped in the other's. Both looked at me anxiously, as if I had news.

"I'm the chaplain," I said quietly.

"Oh, no. Oh, it can't be," one woman said, taking a deep breath. She was a dark-haired, middle-aged woman with the deep olive skin of a Mediterranean. Her eyes were puffy but open wide, as if she expected to hear the worst news possible of her husband's condition.

"I have no news for you," I said quickly. "The doctors are still working on your husband. I thought I might come and stay with you a while." Relief fell across her face as I sat down opposite them.

"What happened?" I inquired.

"Jimmy was gardening," the neighbor started. "I guess he had a phone call. So Lil came out to get him. She walked around the house and saw him keeled over on the lawn. He was grabbing his chest. She screamed so loud I couldn't miss it. So I came running out of my house. When I saw Jimmy on the ground like that, I knew something serious had happened. I called the ambulance right away."

Lil was staring at the floor; her eyes had a faraway, blank look.

"I called her sister," the neighbor continued. "They should be coming soon. Jimmy and Lil have a big family. We don't know them real well, you know how that is. But they're great neighbors, so kind and sweet. We've been next door to them since 1973. There's never been a harsh word between us."

"This isn't happening," Lil said slowly. "What are we supposed to do here? We can't just sit here and do nothing." She sat up, determined, and looked straight ahead with those same glazed eyes.

"We've got to send good vibes to him right now. What do you think? Let's say positive things." She closed her eyes tightly and began repeating positive phrases: "You are going to be better. Your heart is healthy." She wanted to believe these vibes were leaving our little room, traveling down the corridor, entering room one, and putting an end to this nonsense. Lil pleaded with her neighbor to help. "Come on! We've got to do something. I can't do this alone. Jimmy needs us, so just repeat after me—," but it just made the neighbor cry and embrace Lil all the more.

"He's not going to die," Lil said resolutely. She looked at me, her eyes still far away. A chill went down my spine. I had seen that look before. As far as I could tell, Lil was not with us mentally.

A nurse appeared in the doorway. She seemed optimistic.

"The doctors are still working on your husband. As soon as they're able, they will come back and talk to you. In the meantime, is there anything I can get for you? Would you like some water?"

"Is his heart going?" the neighbor asked.

"He is still having arrhythmia, but the doctors are doing everything possible."

Lil made no response. The room fell quiet. The nurse gave a parting word of comfort and left. With that, family members began to arrive. I have always admired the ability of some families to express their emotions freely. Lil's family entered the room crying, with arms opened wide to embrace her. They stood her up and hugged and kissed her, some with loud sobs of unbelief, "Oh, poor Jimmy!" Within fifteen minutes, there must have been thirty family members around the small room.

With each embrace, Lil became more and more distant.

"Why are you here?" she asked in a low monotone. "There's no reason for you to be here. You should go home. Jimmy's going to be fine. Stop crying. There's nothing wrong with him."

The family's presence and robust crying were forcing Lil to face the reality of the crisis. Her young husband was having a heart attack. Doctors were trying to save his life. He might be dying. She might never see him again. Those were the facts, but she refused to face them. Lil's resolve grew deeper, more assured: This was not real; it wasn't happening. There was no crisis. In front of my eyes, I watched her enter a world of denial and protection where none of this could touch her.

She was escaping—to somewhere safely outside reality.

Eventually, her sisters saw the distant look in her eyes—the cold, motionless face responding but not understanding, answering as if from far away. They sat her down and drew her close.

I had never seen it this bad—not like Lil—and it scared me. Deep inside, I understood it. She had no place in her thinking for such an event. She had created a philosophy of life in which suffering didn't happen to her. Sure, it was in the morning papers and on the nightly news. From time to time, friends and people at work had told her stories of others who suffered great tragedies. But it never came close to her.

This is the safe world of denial. We who've been there know it. There's no preparation for the possibility of trauma. We are escapists—living life full of hope and with the promised guarantee that "my family and I" will have long, healthy lives free from trouble. It never dawns on us to think, "What if my Jimmy were to die?" Denial doesn't think seriously about death. We don't make out our wills. We stay away from doctors and hospitals—even when the pain is excruciating. We turn away and make believe.

I remember my cousin wanting to play hide-and-seek when she was a little child. I sat at the breakfast table and told her I would count to ten. At ten, I'd come and find her. I started counting, and she didn't move. When I reached ten, she pulled an amazing trick on me. She closed her eyes. In her three-year-old mind, she had disappeared. She couldn't see me; therefore, she reasoned, I couldn't see her either. I called for her, "Where are you?" But she remained still, without a peep.

Standing there with her eyes shut, she had found a safe hiding place. But I knew where she was.

And I knew Lil was in the same place. I'd been there most of my life. But I never knew the strength or depth of denial until I saw Lil that day. Everyone in the room knew the seriousness of Jimmy's condition. Everyone was experiencing the shock, the grief, and the pain of the tragedy. But not Lil. In spite of all the sounds of wailing, the tears, the embraces, and the whispers in her ear, "I'm so sorry, Lil," she wasn't crying. She had her eyes closed. In her mind, her husband would soon come through the doors of the waiting room, healthy and fine, and prove everyone wrong.

Then she'd open her eyes and come out of hiding—and the game would be over.

The Doctors Arrive

Two doctors weaved their way between the family members standing outside the waiting room. They entered, looked around, and asked for Lil Zampola. One of her sisters said, "She's right here."

"Mrs. Zampola," the older doctor began. He didn't have eye contact. Lil kept staring hard at the floor. "Your husband arrived here with a cardiac arrest. We did everything we could. But his heart was just too weak. It didn't respond to treatment. We lost him about five minutes ago, and I'm very, very sorry."

The place erupted with emotion. People cried out, sobbing, "Jimmy...Jimmy...Jimmy." Lil's sisters began to rock in their chairs, weeping—making Lil rock, too. But she was unchanged. The doctors saw Lil's lack of response. They looked at me and I shook my head, as if to tell them she wasn't able to handle this news. One doctor stooped down to talk with her. After a few minutes, he prescribed a sedative.

My own tears came. The bursting dam of emotion in everyone else overwhelmed me, and seeing poor Lil was like seeing a mirror to my own soul. How I remembered this grief. I had placed my heart in the hands of people I loved, and beyond my control, they were taken away. I didn't want to believe it. I wanted to live life as if it had never happened. But eventually, the loneliness, the emptiness came. I felt it again.

I slipped out of the waiting room and called the family's priest. A few minutes later, the nurse arrived to invite the family to see the dead body. Lil stood up with her sisters and led us down the corridor into room one. Upon entering, the family began its loud lamenting as they saw their beloved Jimmy lying on the table. All the intravenous lines were gone; the monitors had been turned off and stored neatly overhead. Jimmy was dressed in a clean, well-pressed, dark blue-and-white striped hospital gown. A linen bed sheet came up to his waist. His body lay quiet—frozen—as if the violent assault from his heart and the intrusion of the electric paddles and long, sharp needles from the doctors just moments before had never happened.

The family gathered around the bed. Lil stood at her husband's left side, close to his head. One of the doctors stayed by her, watching her carefully. Some of the relatives asked him if he'd explain what happened again. And he did, telling how Jimmy's heart had failed and what they did to try to save him. When he had finished, he looked uncomfortably around the room at the mourners—and then at his watch.

There was nothing more he could do.

The door had no more than closed behind him when Lil seemed to awaken from her stupor. "Look at what you're doing," she said to Jimmy in a stern, angry voice. "You're making all these people upset." With both hands, she pushed hard against his shoulder. One, two, three, four times—making his large body jump from right to left as if with life. It made the younger people shudder with disgust.

"Get up! Get up! Get up! Get up!" she demanded. "Stop this play acting. You're always playing a practical joke and up to no good. This isn't funny anymore. You're wasting everybody's time. When are you going to stop this, Jimmy?" She tried to shake him again, but her sisters restrained her.

"Lil, Jimmy's dead. Look at him, dear," one sister said softly, trying to calm her.

"Did you hear that, Jimmy?" Lil hollered. "You've fooled your own sister-in-law. Now stop the charade. It's time to go home and have your dinner." And on she went, mad that Jimmy had selfishly bothered all these doctors and nurses, plus his own family, with his stupid tricks. She saw his still, pale face. She felt his body turning colder with each passing minute. She heard the unending cries of her family and the echo of the doctor's solemn words: "We lost him." But even then, Lil couldn't see it.

She shook him again. *My husband is not dead.*

I stood at the foot of the bed, my eyes fixed on Lil. I tried to imagine where she was—dug deep into the trenches, on the front lines, trying to wage a war between her reality and ours. She was fighting to win with all she had. Meanwhile, the family's crying kept rising and ebbing as they faced the fact: Jimmy was gone, just like that. The door would swing open, another family member would appear, and the cries would go up once more. His death was like a sword that slashed their own flesh. Standing with them in their shock and grief, I felt awkward. I didn't know what to do. Should I step in, lead them in prayer, and offer words of comfort at this miserable time? Or should I wait for their minister to come? It was best to wait, I thought. They needed their pastor now—someone they knew, loved, and trusted.

His position would allow him the right to join them in their pain. He would pray, pleading with the Lord to shower His mercy and grace upon them at their hour of need. He would serve—by attending the wake, preparing and officiating the funeral service, and leading the family through the hard work of burying Jimmy. Then he'd be there for them in the difficult days ahead. How Lil needed the love of God to gently guide her out of this blinding denial. She needed to face the truth: Her husband was dead, and that's the real world...fallen and full of tragedy, sorrow, and heart attacks. That, for Lil, would take time.

And more, Lil's minister could help her embrace the great promises God has given in the Bible. In her grief and loneliness, she'd find a lasting comfort for her soul that can only come from above.

> I will lift up my eyes to the mountains;
> From whence shall my help come?
> My help comes from the Lord,
> Who made heaven and earth (Psalm 121:1-2).

"Oh, Lord, send their priest," I prayed.

The Pastor's World

"In the Name of the Father, Son, and Holy Spirit."

The door swung open. In that second, the wailing stopped. Those in one another's embrace broke free and stood on their own. In unison, each person made the sign of the cross, reaching to the forehead, the stomach, the left shoulder, the right shoulder, then the heart. I backed up three or four feet, and the priest rightfully took my position at the foot of the bed.

His long, dark brown robe reached to his ankles, where his bare feet were strapped into wide leather sandals. A white rope circled his waist twice and hung down his right side to the knees. His removal of the robe's hood upon entering the room revealed a balding man in his mid-fifties. His tall, stocky frame gave him a certain presence. Using both hands, he clutched a book, holding it a short distance from his face. A pair of bifocals sat upon the middle of his nose. His eyes never strayed once from the words on the page. Not once.

The service of last rites had begun.

Every head was bowed. The family members held their hands at their sides, and closed their eyes. The tears were gone. Even Lil stood at attention, fully participating in this familiar mass. She was suddenly present and conscious.

The priest read quickly, without emotion—first in English, then in Latin. His tone of voice was flat until he came to the close of a prayer. His voice would go up, then down with the "Amen." For fifteen minutes, this went on. When the service demanded a response from the family, they knew every word. They knew when to come in. They spoke with one voice, just as fast and in the same tone of voice. They made the sign of the cross a dozen times. At no point did anyone look at the dead body, break into tears, or lean helplessly into the arms of a loved one. Not during the mass.

At one point, the priest reached into his pocket and pulled out a short staff that looked like a microphone. Three times he shook it at the dead body. Water came out and sprinkled the white sheet in front of him. Then he shook the staff at the family—once to the left, then to the right. Not once did he look up.

The service went on. I watched the priest carefully. I kept wondering, *Did he know Jimmy? Was he surprised by his death? Did it grieve him? Was he suffering from this loss too? And what did he think about Lil? Was he aware of her condition? Had he asked the doctors before coming into the room how she was taking this devastating news? Did he realize how urgently she needed her pastor?*

I kept waiting for him to look at the dead body or direct his attention to Lil—or to any one of these members of his flock. Why wasn't he looking at them? Sometimes the greatest comfort anyone can give is just a simple look of compassion—those gentle, watering eyes of someone we know and trust...eyes that understand, that feel our hurt and bear it with us...eyes that tell us we're loved.

Nothing.

I began to criticize myself. *I shouldn't be so judgmental. This is their tradition.* In one second's time, the chaos of the family's unchecked emotion had come into immediate order. Even Lil, who had built her own safe, little paradise where death isn't real, became alert, clear, focused, and—I suppose—normal. I tried to accept this at face value and not ask too many questions.

But the fact is, I didn't understand what was going on. Why was the family so separated from their grief during the service? Were they not allowed to express their sorrow in worship to the Lord? Was there something wrong with crying in front of the priest or before God? Surely, that wasn't right. The Lord doesn't remove Himself from our pain. His promise is to be with us as our help, our strength. Didn't they know that? Or did they think God was distant, uncaring, and cold to their suffering?

These questions kept rolling over and over in my mind. I

wanted to know why they were hiding their true feelings—and why the priest was so formal and removed. I finally reasoned that when the service ended, the priest would go over to Lil and hold her. Then he'd remain with the family, go back to their home, and stay with them in these first hours without Jimmy. That's what would happen, I was convinced. He'd be there, fully present to extend the love and peace of Jesus Christ in their misery.

I took comfort in these thoughts and bowed my head to join them in their prayers.

The service ended a few minutes later with a curt but vigorous "Amen" from the priest. He closed the book and turned ninety degrees to the right, facing the door. His eyes never met the widow, a family member, or the dead body of Jimmy. His work was done. He pulled his hood over his head and walked to the door. It opened. It closed. As soon as it shut, the wailing began all over again.

A Place Called "Safe"

Why didn't I do that? I wondered. I could have arrived at the hospital, prayed with Lil in the waiting room, called their priest, and gone home to finish the lawn. Why did I stay? *Why choose to enter someone else's suffering?* Jimmy's death had now impacted me, and I didn't even know him. I drove home grieving for Lil, the family, and even for the priest. In my mind, I kept seeing Lil rock her husband's body, denying his death. It brought back a rush of memories from my past. Who needed that?

I shouldn't have stayed. I should have kept my distance.

But that's not the answer, and I knew it was wrong. I had a choice, as if two roads were diverging in front of me. One road was easy, chock full of denial and a continuation of my past: *Don't enter their suffering; act as if it's not there.* The other was new and more challenging: *Stay with them and be there for them; face the facts.*

The first option was most familiar. There's a game children play called "tag." A player is "it." His job is to tag someone else, and that person becomes "it." The object of the game is to never get caught, never be "it." The game has a place, like a tree or a lamppost, called "safe." As long as a person stays there, the person who is "it" can't touch them. But the moment they leave, they're fair game to be caught.

It's a simple game, but growing up I found that it held the secret to life.

I believed there was a real place in the world called "safe," where nothing could touch me. No sickness, tragedy, suffering, sorrow, tears, or death. As long as I didn't leave "safe," "it" couldn't come too close. Not to me! Others got caught because they took risks, but I chose to stay at "safe."

It's one aspect of denial: *Pretend danger is not there!* I grew up safe and protected, holding tight to everything that was safe, believing nothing could harm me. But this view of life failed me, as it also failed Lil. In the bat of an eye, she found herself at the Stamford Hospital emergency room with her husband in a cardiac arrest. Just like that, she was far away from "safe," as if it had disappeared into thin air. She was suddenly vulnerable, alone, helpless, and face to face with "it"—in this case, death. Her "safe" was gone. She was now forced to meet suffering, whether she wanted to or not. That's life, and she wasn't ready.

It's a hostile, unfair world when the rules change without notice. "Safe" is supposed to be safe, both now and always. But real life isn't like that. There's no place on earth called "safe." Lil found that out the hard way. And even then, she couldn't face the suffering. She ran, trying to find another "safe."

A second aspect of denial is *positive thinking*. As long as we hold onto "safe," this psychology works. Danger can come our way. It can scare us with a roar; it can jump up and down and paint a frightening picture of our future. But positive thinking says, *All we need to do is remember we're "safe."* As long as we stay, nothing bad can touch us. Therefore, we're to think positive thoughts, see life optimistically, and be free of negative influences. "It" has no power. If we stop looking at "it," we'll be successful and happy.

This popular dime store psychology fails the test in crisis. What happens when "safe" disappears and there's no place to hide? What happens when tragedy comes in all its terrifying power, as it did to Lil? Are we to think positively when nothing is positive? But how can we do that when we're hurting, lost, grieving, helpless, sad, lonely, or wholly uncertain of ourselves and our future? Death is like that, and so is abuse, rape, depression, and mental illness. To think positively is to deny the seriousness of the tragedy.

We need real answers to real problems, not philosophies and psychologies that deny the reality of life. We need answers that face the fact that crisis and suffering come without notice. Positive thinking cannot change that, nor can it help us through our dark, powerless days. Lil tried positive thinking in the waiting room. She sent positive thoughts to her husband, hoping they'd bring him back to health. But all her positive thoughts couldn't stop what was happening in room one. Jimmy was dying, and nothing about it was positive. As hard as she tried, she couldn't stop him from dying. In the end, this psychology was nothing less than her attempt to find another "safe" world where bad things don't happen.

Cheap psychology only sells when life is rosy, not when it stinks.

There is a third aspect of denial: *Work hard and fill up the time!* Denial is just as powerful after the tragedy. When Mom died, I immersed myself in high school activities—classes, sports, band, theater, and friends. I went to college and filled my schedule with classes, took summer courses, and graduated as fast as I could. I was running from my grief. I was afraid to stand still and face what was inside.

Combining long hours of work with a hectic social calendar

is a common way to hide. Of course, I knew Mom had died. But I worked hard at creating a life for myself in which I couldn't feel the loss, nor did I have any time to think about it. The energy to grieve became the energy to live a busy life. I kept my focus: I was going to be an Episcopal priest. I kept going, hard and strong, my eyes fixed on that goal.

But if I had stayed in this denial, what kind of priest would I have become? How would I have handled Jimmy's death? Would I have come into the hospital room never looking at the dead body, never acknowledging the widow, and never hearing the deep, wailing cries of the family? Would I, like Lil's priest, enter the room, do my work, perform the service, and then—just as I've always dealt with human suffering—escape as fast as I could?

This priest did his work, but he never stepped into the pain. He stayed at "safe," where Jimmy's death couldn't touch him. And I understood that. But was it right and fair to the family, his own church members? They were suffering, out in the cold, far from that place called "safe," vulnerable and unprotected. The priest had made a choice. He decided to be so immersed in his work that there would be no time to deal with the pain.

Three aspects of denial, and I was familiar with them all. Like Lil, I could close my eyes and pretend danger wasn't there. I could embrace the positive thinking mind-set and say, "Everything will be fine!" And when crisis came, I could fill my time with work and things to do. These were three ways of staying as safe as possible.

If I had continued on this path, I'd have become like the priest who couldn't stay with the family that day in the hospital. Denial would have remained my philosophy of life, and I too would have made the easy choice: Never enter into suffering—not your own or someone else's. Always stay at "safe." Don't let "it" come near.

Is that how God is? As a priest, if I modeled that view of life, wouldn't I be telling my church and the world around me that God is the same way? Does He stay outside, never entering our suffering? Does He expect us to stop crying when He enters the room and say our mass without emotion or all the feelings of loss and grief? And then, after His blessing, does He leave us to experience the suffering and sadness of death on our own, without Him? Is that the God of the Bible whom we know and serve?

From the beginning, denial has never been God's way. Never. Then why should it be mine?

Staying "safe"—that was me. Driving home from Stamford Hospital that night, I realized I had made a choice contrary to my past. I had stepped out of "safe" and stayed with the family in their suffering. I didn't like it. I was grieving and it hurt. But observing Lil and her priest scared me. I didn't want to take the easy road, lost in denial. It doesn't work. It doesn't answer our deepest questions.

It's a choice we make—and a costly one—to willingly enter a world of pain and sorrow.

Making the Choice

A memory came to me.

It was my first day of school. I had on a new outfit and carried a new lunch pail and a clean notebook. I had a sharp, unused pencil in my shirt pocket. It was a big day. My older brother and sister were in grades two and four. They had always left me behind. Now I was going to school, just like them. I was growing up and nervously excited about this new adventure.

Gregor and Kate had already gone when my ride pulled into the driveway and honked. I looked out the window and then turned to Mom. She bent down and hugged me. "Oh, I wish my little boy wasn't growing up." She smiled softly, stood up, and opened the door.

I remember looking up at her, tears filling my eyes, and saying, "I don't want to go." The desire to be like my brother and sister was gone. I wanted to stay home. I didn't want to grow up. I didn't want to face the world out there. I didn't like the feeling of loss.

She stood at the door. It was time to go.

A thousand times since that day, I have stood at doors like that and faced the same decision: To stay is safe. To cross is unsafe and unknown. It's a journey that may lead me far from home. I had to go that first day. I had no choice. But many times since then, when the choice was mine, I have stood at the door and not crossed over the threshold. I've chosen to stay safe, to remain at home, to freeze time and remain Mom's "little boy."

How long would I continue to make such choices? When would I learn that time can't be frozen and safety can't be found in this world? What would it take to choose the risk, to step into the unknown, to confront the suffering and stay there?

The door was opened. The choice to cross over was still mine.

Chapter Three

FRIDAY MORNING

The morning after the plane crash Erilynne and I were sitting at the breakfast table when the phone rang.

"Thad, it's Alden Hathaway. I'm coming out to see you."

"I'd like that. How about I meet you at the church around 8:30?"

"See you then," my bishop said and hung up.

I went upstairs and knocked on the guest room door. John and Harriet Rucyahana from Uganda, East Africa, were staying with us during their three-month preaching mission to the United States. I told John about the bishop's visit. He was anxious to come with us. Then I called Ken Ross on the phone and asked him to meet us at the church. The bishop's call set our morning in motion. I had wondered during the night what we might be called on to do in this emergency. Did the rescue operation have a place for the clergy?

The phone rang again. Erilynne took the call. It was Bob Dwyer, our friend at USAir systems control. He was helping to form a team of clergy and trained counselors to meet with the families of Flight 427 passengers. USAir was flying family members to Pittsburgh from all over the country. They had secured a banquet hall at the Pittsburgh Airport Marriott to receive them. USAir wanted to offer as much support as possible.

"Why don't I go to the Marriott" Erilynne said to me with Bob still on the line. "You can go meet with Bishop Hathaway and maybe take him to the Plaza." It seemed like the right plan.

John and I met Ken at the church. None of us were sure why the bishop was coming or what he wanted to do. As we waited for him, the church phone rang.

"Thad, it's David from New Hampshire." The Rev. David P. Jones had been the bishop's assistant for many years before taking a large church in Concord. He had been instrumental in helping us start our church mission. "You've got to know we're praying for you," he said. "I'm sure Christians all over the country are praying for you and all the other clergy assisting in this awful tragedy." We talked awhile. I thanked him for his call and concern. I couldn't help the sudden feeling of restlessness and fear that came over me. It was as if David knew what was going to be required of me—*and I didn't.* I hadn't thought about it until that moment. My mind raced. What was ahead? It felt as if I was stepping deep into unknown waters.

The bishop appeared in our office and greeted us warmly.

"I'm not exactly sure what we should do," he said. "I've no specific agenda. I woke up compelled to be with you this morning, to support you in your ministry at the crash site, and to pray with you."

"Bishop, I was thinking," I jumped in, "we should go to the Green Garden Plaza. We can show you what happened last night and maybe get a feel for what's going on now."

"Let's do it," he said. Our group left the church in two cars so that Bishop Hathaway could leave for his late morning appointments; I went with the bishop. We drove the winding country road just as I'd done the night before, and I described for him the events following the crash. Once again, the Hopewell police were stationed at the entrance of the Plaza, questioning each driver. The bishop rolled down his window, saying "Episcopal Bishop of Pittsburgh. The next car is with me, too." We were waved in.

The Plaza looked so different from the night before. All the emergency vehicles were gone—no ambulances, no fire trucks. Now the parking lot was a circus of cars, Salvation Army and Red Cross trucks, and all kinds of rescue equipment. Vans and a sea of trailers were being used for office space, supply storage, and canteens much like the mobile cafeterias used by the military. County buses were scattered throughout the Plaza, ready to transport rescue workers to the site. To our left as we entered the Plaza was a large, blue-and-white-striped pup tent. I'd seen it on the early morning news. It was being used for press conferences.

Behind the pup tent and near the side entrance of the Unis car dealership was a mass of people, cameras, and lights. Farther back I saw huge trucks with satellite dishes on top and the familiar symbols of ABC, CBS, NBC, and CNN. It was hard to believe that the international press community had descended on our small town. It was even harder to believe that a commercial jetliner had crashed in our backyard. One day later, and the news was just as shocking.

We found a place to park and got out in a steady, light rain. Where do we go? Who do we see? How do we get involved? What do clergy do at times like these? There was no reason to stay still. The bishop, John, Ken, and I decided the best thing to do was to find the emergency command headquarters. They would know what was going on and where we could best serve.

The Salvation Army

On the way, Bishop Hathaway spotted the red insignia for the Salvation Army. He felt we should stop and offer our support. A number of their men, dressed in formal, dark blue, military-like uniforms, were standing outside the Salvation Army trailer talking with people. Major Robert W. Pfeiffer saw us from inside the trailer and came out.

"Bishop Hathaway, I'm Bob Pfeiffer. We did a service together a few years ago in Aliquippa." They shook hands, remembering their first meeting. The bishop then introduced us to the major.

"I'm sorry we're meeting under such terrible circumstances," Bob said with his distinctive Boston accent. "This is a great tragedy. One hundred and thirty-two people died last night just flying home from Chicago."

"What's going on now, Bob?" the bishop asked.

"Well, the crash site is closed until early afternoon. The National Guard is up there. They're staking out the boundaries. It's not a big crash site. It's not like Lockerbie, Scotland, where the plane exploded thousands of feet in the air. That crash site was miles wide and miles long. This plane came straight in. Also, a carefully designed grid is being set up so that everything removed from the site can be tagged with a number. This will help the National Transportation Safety Board figure out how the plane came in and what happened on impact. It may even give them clues as to what caused the crash."

"They still don't know?" I wondered.

"No, everybody has a different theory. The Federal Aviation Administration (FAA) and the National Transportation Safety Board (NTSB) have secured the flight data recorder, but they aren't saying anything. It's too early to know. And they're anxious to get up the hill. When the grid work is done, the site will be opened and rescue workers will be allowed in. They should find some answers in all that rubble."

"Is there easy access to the site?" Ken asked.

"Not really," Bob answered. "There's an old logging road a short distance from where the nose of the plane hit the ground, but it can't handle the kind of equipment that has to get back there. There's a group up there now called the Delta Team, I think from Allegheny County. They're the best. They move at lightning speeds and have promised two access points into the site by the end of the day."

"Bob, is there any sense of order down here at the Plaza?" the bishop said, looking around.

"Oh, yes. There's a master plan in effect for the rescue operation of a plane crash. Everyone has his or her specific jurisdiction—the FAA, NTSB, county coroner, county and state officials, the Hopewell police, and even the Salvation Army. I know it looks confusing down here, but everyone has his work to do and everyone has a boss somewhere. Look over there," he said, pointing to a large white trailer with a sign over its door. "That's the emergency command headquarters. Everything starts and ends at that office."

"And what's your particular job?" the bishop continued.

"I'm basically overseeing our rescue effort for Flight 427 at the Plaza. The Salvation Army helps people in times of disaster. Our canteen was already up at the site two hours after the plane crashed. The rest of our emergency equipment in Pittsburgh arrived through the night. We're providing food for the rescue workers at the site, the Plaza, and eventually, at the morgue. We've got trained counselors ready to minister to the workers when they come down from the site, and to anybody else who's suffering from this tragedy. I'm also helping out with the press and community relations. We're here to serve. Already, we've made a dozen or more trips to the crash site. Officials needed rides, and we're here to help."

"You've been up there?" John came in quickly.

"Oh, it's an awful sight," the major said, looking down and shaking his head. "I've never seen anything like it, and I hope I never do again."

Bob's words stopped the conversation. He had been up there. He had experienced the site firsthand. All our questions about procedure and the morning's activities were dwarfed in comparison with the anguish on his face as he remembered what he'd seen. I caught my breath. I couldn't imagine it—going to the site! That was something for trained experts, not the clergy. The thought of him going, of him physically standing at the crash site, was beyond me. It horrified me. How did he do it? Why did he do it? I felt myself take a step backwards, almost defensively. It came too close. He went, but I couldn't. I admired Bob's courage. I was impressed with the mission of the Salvation Army to go to the front lines, set up their equipment, and serve the rescue workers. But no way was this something I'd do. Not a chance!

It was time to change the subject. We still didn't have an answer to our question: Where could we best serve? Did the Salvation Army need assistance? Should we go to the emergency command headquarters? Before I could ask, the major turned to Bishop Hathaway with a question of his own.

"You know, Bishop, I think it might be a good idea for you and your men to go to the crash site. If you're going to minister to the families and the workers, you need to see it for yourself."

"You can take us up?" he asked.

"We have a truck right here." Bob turned to one of his men and asked which truck was available and who could drive us. I couldn't believe my ears. There was no way I was going up there. How would our being there help minister to the families and the workers? I looked at the bishop, John, and then Ken. I couldn't say anything. It felt as if every ounce of strength had left me—a horrid feeling—weak in the knees, dry mouth, racing heart, sweaty palms. Go to the site? *Me?* The bishop had to refuse this offer.

"I've got a man and a truck ready to go, Bishop," Bob reported.

"It's important that we go," the bishop answered.

To the Crash Site

On the drive up, I couldn't focus. One minute, I was thinking only about me: *What have I gotten myself into?* I'd never been to war, never seen violent death, and never been tested on what I could and could not handle. The next minute, I thought about the rescue workers. How many of them were prepared to work at the crash site? Were they trained for something of this magnitude? Then, the next minute, I thought of the passengers and crew of Flight 427 and their last twenty-three seconds spiraling to earth.

The rain was still coming down.

The driver, Mark, and the bishop were in the front, talking. I said to John and Ken, "I've never done this before." Ken shook his head. He hadn't either. He looked as solemn and uneasy as I felt.

"You never get used to it," John said. "Never." He had seen it all before. He had been in charge of St. Peter's Cathedral in Hoima, Uganda, during Idi Amin's reign. He was no stranger to gunfire, mass killings, and rescuing dead bodies from the main road of town—at the risk of his own life—for proper burial. He had even stood before Idi Amin's soldiers with a machine gun thrust at his temple. John knew about violent death.

We held hands together and prayed. I drew strength from John. He too was downcast and solemn, as if it didn't matter that this wouldn't be his first time. It seemed that the grace of the Lord was required just as much the hundredth time as it was the first. I needed that grace; I begged God for it. I was scared.

Mark took the main road up the hill by the Plaza. Halfway up, he pointed to a road on the right. "There's Beta," he said in a loud voice. "When the Delta Team finishes their work tonight, that road will be connected to the crash site. All the plane parts will be decontaminated on the Beta side and then taken out from that road." He continued driving up the hill. Near the top, three policeman guarded the entrance to a dirt road on our right. Mark waved without stopping and made the right turn as one of the policemen waved back. "There's the Delta Team," he said, pointing straight ahead to a group of men on bulldozers forging a new road. We followed along, bumping and jerking here and there.

"We're coming in on the Alpha side," Mark continued. "That's where our Salvation Army canteen is set up. Rescue workers can enter the site from either Alpha or Beta. But they must leave through Alpha. That's where they'll be decontaminated and the buses will take them back to the Plaza."

"What does that mean?" the bishop asked.

"Every rescue worker will be fitted for a decontamination suit at the Plaza before coming to the crash site. When their work is done, their suits come off, and they'll be washed down."

"You know," John said to Ken and me privately, "this isn't like my experience in Africa. The bloodshed from Idi Amin and Milton Obote, his successor, was different. Or the genocide I saw in Rwanda this summer, with the mass graves and church floors and walls covered with bullet holes and blood. War is different. You feel the oppression of hatred and murder. You know this is what man has done to man. In war, you feel the blackness, the demonic, as if you are walking through hell itself.

"What we're about to see," he continued, "was an accident. It may look the same, but there is comfort in knowing that this terrible human tragedy was not caused by greed or jealousy or hatred."

I understood his point. He was right, and that did make a difference. But it didn't stop the feeling of butterflies deep in the pit of my stomach. It was all beyond me—war or accident.

The road turned left into an open space, where the Salvation Army canteen was parked. The driver pulled up next to it and stopped. "Stay here. Let me find out what's going on." He got out and walked over to the canteen. A man dressed in a similar uniform came out to meet him.

"Look at that," the bishop said sadly. The terrain in front of us sloped down into a valley and then turned up. We were still far away, but from this distance we could see debris scattered all over the hillside. "I expected to see something of the plane," the bishop whispered. "It's just not there."

Mark motioned for us to get out.

"There's a yellow tape stretching over the logging road," he said. "See it?" The dirt and gravel road winded down five hundred feet and flattened. Just as it began to head up, a yellow tape stretched across it. "That's where the crash site begins. We're not allowed to go beyond that. Also, the word is that we can't stay long. The county coroner has asked all nonessential personnel to leave the site."

"Can we go to the tape and have some time for prayer?" the bishop asked.

"I think that would be fine," Mark answered.

We began walking down the logging road. Off to our left stood two dozen men and women of the National Guard. Each was dressed in military boots, a camouflage green and black jumpsuit with a matching poncho, and a wide, black belt at the waist. One man stood in front of them giving detailed orders in a calm, clear voice.

The road hugged the side of the hill. Up ahead and to our left, the land shot straight down into a ravine riddled with wreckage. To our right, the hill went up. Straight ahead, a hundred feet past the yellow tape and a short distance to the right of the logging path, I saw a large silver sheet of metal, maybe ten feet wide by five feet high, with the familiar red and blue USAir stripes.

"Look, there's part of the fuselage," I said, pointing at it.

"They say that's one of the biggest pieces of the plane," Mark came back.

"Biggest pieces?" the bishop said in amazement.

"There's not much left."

The closer we got to the yellow tape, the more came into view and the more we felt the power of the destruction, the force of the explosion. There was nothing left untouched. All was demolished...all ravaged as if an earthquake had come, a bomb had exploded, or a tornado had touched down and refused to go back up. Little was immediately recognizable. But each step closer meant more definition, more feeling of loss, human suffering, and death—and the more I wanted to turn and run away.

We stopped at the tape.

Prayer at the Yellow Tape

Up the logging road I saw a group of men walking slowly toward us. One of them was tall and slender, his arm bound in a sling. I knew that was Wayne Tatalovich. My eyes roamed, taking in some clothing in the trees, a man's torn brown shoe, the small pieces of twisted metal blackened by the fire, the layers of printed paper from a weekly magazine.

"Oh, Lord, bless the family of this dear woman," the bishop prayed quietly. To our right were human remains. They had no specific shape. But a woman's tan leather pocketbook lay next to the body, which was broken and charred. In front of us, some ten yards away, was the body of a man. I tried to take my eyes away and I couldn't. There was shape. His hand lay atop his body.

"It's beyond words" Wayne said as he passed by the dead man. Wayne's face was tired and agonized, but his eyes were focused, in control. "I've been in the funeral business for thirty-four years, and I've never seen such devastation." I introduced him to the bishop and John.

"Did you sleep at all last night?" I asked.

"We got to bed around two, but we didn't sleep. How could anybody sleep? You know, Val and I were at the golf club a half mile from here. We heard the explosion, saw the smoke. We looked at each other in absolute disbelief. We knew exactly what had happened. It took about ten minutes to get here."

"This isn't going to be an easy job," the bishop commented.

"If we're able to identify 25 percent of the bodies, I'll be surprised." Wayne's portable phone rang. He excused himself and began to walk away. Seconds later he turned back to us. "I've got to go. Good to meet you," he said to the bishop and John. He crossed over the yellow tape and looked at Ken and me. "Thanks for the prayer you guys left last night. It meant a lot to both of us. We read it again this morning." He gave a half smile, turned to join the other men, and walked up toward the canteen.

For a few minutes, we stood in silence at the tape, facing the crash site. The longer we stayed, the more difficult it became. It was as if my eyes were suddenly opened to see that there were human remains everywhere. I bowed my head and closed my eyes hard. The rain made me cold, and so did the fear. I had seen too much. Just past this yellow tape was a place that scared me. It was as if all my childhood nightmares had come to life in a single moment and were rising up to frighten me. I used to scream, wake up, and find myself at home—safe in the arms of those who loved me.

But there was nothing safe here.

I couldn't help but think of the rescue workers. How were they going to step over this yellow tape? Who is the person who can see such violence, such human suffering, and not feel the fire of burning torture to the heart, mind, and soul? Can people be trained to see this mechanically, as if it were another day's work? It can't be. Like John said, "You never get used to it. Never." Those workers were coming with hearts just as vulnerable, just as scared as mine. *But they'll have to cross the tape,* I thought. *It's their job to enter that nightmare world out there.*

Just then, the bishop raised his right hand, closed his eyes, and began to pray. "Bless, O Lord, God and Father Almighty, this land of death and destruction. Bless each person who works at this crash site. Give them Your heavenly grace that they may have the strength to carry out their tasks and so care for the dead. Let those who cross over this yellow tape know Thy presence and Thy promise never to leave us or forsake us...."

On he went, praying for the rescue workers. In my mind, I kept seeing them. I didn't understand how they were going to handle this experience. If Wayne Tatalovich had never seen anything like this, then neither had these workers. And, I wondered, how many would come without a living faith in Jesus Christ? We had come praying, asking for the Lord's grace and strength. I couldn't imagine being here without having Him in my life. Where would these men and women turn for comfort when they saw violent death as they had never seen it before? Where would they run when the darkness of despair set in?

How would they cross the yellow tape?

The bishop prayed for the family members of the 132 victims of Flight 427. He prayed for all those directly affected by the crash, especially our Hopewell community, USAir and airport personnel, friends and fellow workers of the dead, and even the man in the control tower who last spoke to the pilots. Then he prayed the unexpected. With each word, I felt a growing desire to cup my ears and not hear it: "And we ask Thee, O Lord, to send to this crash site Your ministers. Give them, we pray, Your compassion and courage to stand with the rescue workers as they labor in this valley of death. Let them be witnesses of the saving love of our Lord Jesus Christ, who bore all our suffering at the cross of Calvary, especially to those workers who do not know Thee, Lord. We pray for them especially. And we ask this all in the gracious name of the Father, the Son, and the Holy Spirit. Amen."

The bishop turned and led us back up the road to the truck.

Eagles, they say, teach their young to fly by kicking them out of the nest. Down they fall. Their wings spread, but to no effect. Down to the earth—in a panic, unable to save themselves. They never die this way; instead, the mother eagle flies down, picks them up, and puts them back in the nest until next time.

I hadn't felt that fear and panic of falling since I was twelve. Now it was here again, as if I'd never grown up. It was the bishop's last prayer. He had asked the Lord to send His ministers to the site with the rescue workers. With these eloquent, noble words, I had felt the flick of an eagle's wing. I was out of the nest. Falling.

I knew as the bishop prayed, he was praying for me. Deep in my soul, I knew the Lord's call. I'd be back to that yellow tape; I'd have to step over it and enter that world I had feared all my life.

Out of the Nest

When I was twelve, it was time, my parents thought, for the youngest Barnum to feel his wings. They chose a summer camp and signed me up for four weeks. It was my first time away from home—*ever.*

Dad flew with me to Newark, New Jersey, where the camp director, known as "Chief," met us. I had tried to prevent this moment for months. The closer the time came, the more I refused to go and the more my protests went unheard. "This is part of growing up," I was told. But I knew what it was going to feel like to leave my home—my brother, my sister, my mom, my dad, my room, my dog...my only world. One by one, I'd have to say good-bye as if I was never coming back. At least that's how it felt.

We got my luggage at baggage claim. Dad turned to Chief, shook his hand, said a few words, and then came to me with his arms opened wide; a warm, gentle, loving man. He kissed me, hugged me, and told me I was going to make it. We'd see each other soon. He loved me. I couldn't hold back the tears, the sobbing, the pleading: *Don't leave me here! Not miles from home. Not with Chief. Not ever.*

Dad broke away. He knew what was best for his little boy.

I had never known such pain before. Chief had other little Indians to round up from various flights. We took a shuttle to a nearby hotel. I, like a mule with his back legs dug firm, wanted nothing to do with Chief or his pack. They went to the pool for fun. I stayed in the room, crying uncontrollably. I had never felt this alone—had never known this empty feeling in the pit of my stomach. Like the eaglet dropping from the nest, I had no resources to survive this fall. My wings spread, but to no effect. I had never experienced the fear of falling or the rush of adrenaline. For the first time, I felt that my life was at risk.

Homesickness. The front door closes behind. It's the first step out of the safety of home—warm, familiar, secure—into the unknown outdoors—cold, open, and exposed to the elements.

I went to my duffel bag, opened it, and found a note from Mom. She wanted to surprise me, to assure me that the time would pass quickly—"Just like that! You'll see"—and that she was proud of me. This was a big step. I was becoming a man, and nothing could please her more. She closed the letter, "I love you!" and as I read it, the pain went deeper, the longing became greater, and the absence of love grew more present. I wanted nothing more than to run back home and never leave, never grow up, never become a man.

How could this heartache come from my parents' love? It didn't feel like it that day. But later, when I got my wings and learned to fly, I knew their love and the wisdom of flicking me out of the nest.

What my parents did not teach me was this: Homesickness never really goes away. This earthly life is a never ending cycle that begins with building a home and finding security in those we love, the things we possess, the jobs we hold, the money we make, and the routines we cherish. Then suddenly, we find ourselves falling, as if thrust out of the nest. We go from safe to unsafe—anchored, then adrift at sea.

So, we do it again. We build a new home. It's part of life to keep finding safe places, homes filled with love, people we call family. But time marches on and those we love leave or die—or worse, it comes time for us to die and we must do the leaving. Whoever said, "time heals"? It most certainly does not! Time is the disease, homesickness the symptom. Time, like cancer, eats away at all our earthly homes, robbing us of our loved ones, forcing us to abandon our safe places of rest, making us feel homeless and lost, and demanding we grow up and face the truth about this life.

Nothing here is really safe. No place here is home forever.

The Yellow Tape

On the drive down to the Plaza, I felt the old feelings I had had in the New Jersey hotel when I was just a kid—the adrenaline flowing, the fear of the unknown, the helplessness of falling without any sense of control, and the pain of being left alone, far from home. I didn't want to go back. I didn't want to cross the yellow tape.

How could I take the step and walk into that death and destruction? How could I help others when I myself might not be able to handle such violent suffering? Why did the bishop pray that prayer? *It's not me. I can't do it. I won't do it. There has got to be some other way to serve the Lord in this crisis. There must be other clergy who are be willing to cross that tape, but not me. I don't want to go.* ✓

We pulled into the Plaza parking lot. I remembered my father's embrace at the Newark airport. My mother's letter in the duffel bag. They had such loving confidence that I could step over my twelve-year-old yellow tape and enter that scary, unknown world. And I had none. Now my Father in heaven was calling me to do the same. Just like before.

Step over the yellow tape.

Chapter Four

THE DECISION TO GO

We got out of the truck, thanked Mark, and made our way to the car dealership through a sea of reporters. The bishop needed to call the office and cancel his late morning appointments. By then, the rain had stopped. One half of the sky was dark, threatening more rain. The other half was bright with sun, the blue sky and white, cotton ball-clouds promising hope. *Such indecision,* I thought. *Just like me!* I was torn between going back to the crash site and staying at the Plaza.

Ten feet in front of the showroom, I heard someone call out for the bishop, "Father, can we get a few words from you?" CNN was the first. The bishop's opening statement revealed that we had just been to the site. With that, like sharks to blood, the press was all around us: "What did you see? What's happening up there now? Why is it taking so long to get the rescue teams to the site? What do you have to say to the family members? Why did this happen? Does God cause tragedy? Couldn't He have prevented this plane from crashing? How did you feel up there, seeing the dead?"

Another reporter from CNN came to me, his cameraman right behind him. "Excuse me, do you think family members will be allowed to go to the crash site?"

"No, I don't," I said sharply. "You don't understand. The bodies are not recognizable."

"Does that make it harder for the families?"

"Yes. They won't be able to see their loved ones ever again. Most of us, when we grieve the loss of someone we love, can see them, touch them, and hold their dead body. It makes their death real to us. But not so in this case. The family members of Flight 427 will never see their dead. They'll never have the chance to be with them and say good-bye. That's painfully hard. It makes their death that much more unbelievable. It will take longer for these families to stop expecting their loved ones to come through the door at the end of the day. As horrible as it is to grieve over those we love, it's more tragic when there's no finality, no good-bye. And that's the story here for these families."

I spoke fast and with conviction. I knew the grief of not seeing a loved one who had died and could feel it rising in me. My face and a sound bite of my words went out over CNN news for the next twenty-four hours.

A man from the local ABC affiliate stepped in and asked John, Ken, and me to follow him for a live interview. For the next half hour, we were questioned by a number of television and radio reporters from all over the country. Time and again, we were asked to explain why this plane went down, why people die in tragedies, and why God seems to stand helplessly by. If we went to the crash site for no other reason than to speak for the Lord and His gospel on television, then I say it was worth going.

"Are you an Episcopal priest?" a salesman from the Unis dealership asked me.

"Yes, I am."

"You're wanted on the phone. In the showroom, third office on the left."

I made my way into the car dealership, picked up the phone, and talked with the religion editor of the *Pittsburgh Post-Gazette.* During our conversation, I couldn't help noticing a group of men and women over by the main entrance of the showroom. They formed an uneven, almond-shaped circle and looked like they were talking casually. Each of them was dressed in a long, dark blue jumpsuit with the word "CORONER" written in large yellow letters on the back.

I put the phone down as the bishop came into the room. He still hadn't called his office. I left him alone, wandered into the showroom, and began talking with a man from the emergency command headquarters. He told me that Bob Casey, the governor of Pennsylvania, was coming. In a few minutes, he would enter the back of the dealership and be ushered upstairs to a large room.

"Is it possible for my bishop to greet the governor?" I asked.

"I think it can be arranged," he responded.

The bishop came out of the office and walked toward me. "I'm going to have to go," he said. "I need to stop at the airport Marriott where Erilynne and a number of our clergy are ministering to family members." I told him that the governor was coming. He agreed that a brief meeting was important.

As we waited for Governor Casey, I found myself still staring at the coroners in the middle of the showroom. They were dressed for action and clearly waiting for the word to go to the site. What were they thinking about? Did any of their experiences prepare them for this moment? How were they feeling? Were they confident? Nervous? Unsure? What was it like to be a coroner at a plane crash?

I left the bishop's side, walked over, and introduced myself.

The Coroners

"My name is Thad Barnum. I'm a priest at a local church in Hopewell Township." I put my hand out. A tall, dark-haired man named Mike shook my hand. A few of the fifteen coroners backed away at the site of a clergyman. The rest stayed.

"What's the plan?" I asked.

"We're waiting to go up," Mike replied. "They said about 12:30, but we're not sure."

"I can't imagine doing your work," I blurted out honestly. "But you deal with death every day. That must make it easier for you to go to the crash site, doesn't it?"

"We don't deal with death like this," Sandy responded. She, like most of them, was in her late twenties or early thirties. Her voice sounded troubled, anxious. "I've worked at suicides and car accidents where people were badly mangled. I also worked at a train wreck a few years ago."

"Yeah, I was there," a blond-haired man with a medium build said. "That was the worst I've seen."

"But I've never done anything like this before," Sandy said, shaking her head with dismay.

"None of us have worked at something of this magnitude," said Mike. "The reports I've heard say this plane was just under six thousand feet in the air. For whatever reason, it turned straight down. The engines were at full throttle the whole way. Add gravity force, and this plane came nose in, crashing at three hundred miles an hour. We've heard there's nothing left up there."

"The county coroner told us he'd be surprised if 25 percent of the bodies can be identified," I interjected. That caught their attention—even those who had backed away. "Anyway, we went up and saw the crash site, and you're right, the plane was demolished. Except for one piece of the fuselage, it was hard to even tell it was a plane. We weren't allowed in. They're still defining the perimeters."

"Were the refrigeration trucks up there yet?" one of them asked.

"I don't know what you mean."

"That's where the dead bodies will go. Our job is to find the remains, write down where we found them, and carry them to the refrigeration trucks. From there, they go to the morgue."

"I didn't see them. We were on the Alpha side. Maybe they were on Beta? I don't know."

"Why did you go up?" Mike asked.

"We went there to pray, mainly," I started, "not only for the family members, but for you who are going to work at the site. We didn't think it was right for you to go up there, tell you we were praying for you, and not go see what you're going to see or experience what you're going to experience. That doesn't seem fair, does it?"

"No, it doesn't. Thanks. Thanks for praying," Sandy said quietly. It didn't seem as if my words had comforted her at all. But I understood that. I remembered our ride to the site. John Rucyahana tried to comfort us by sharing his experience with violent death in Africa. I had heard his words, but it didn't relieve the aching, nervous pain inside—the butterflies of fear! Nothing could relieve the tension, save getting out of the truck and going home.

I was impressed with the courage of these men and women. I could see they were restless, apprehensive. But who wouldn't be? How could anyone want to go into that living hell and rummage through heaps of debris, looking for 132 dead bodies? More than ever, I was convinced they needed people praying for them who knew what they were going through. They needed, just as the bishop had said, Christians who would be with them as they sifted through the rubble. They surely didn't need the "warm fuzzies" of clergy who would weep with them at the

Plaza, but not stand with them in that valley of death.

The coroners were going. It was their job. With more conviction, I knew it was mine too.

Governor Casey's arrival was announced. Bishop Hathaway met with him briefly and then led us out toward the car. We lost another twenty minutes to the press, who first snagged the bishop and then the rest of us for more interviews. We finally broke away and headed to the Airport Marriott.

I couldn't wait to talk to Erilynne. I wanted to know how she was doing. What was it like caring for the grieving family members and friends? I wanted to tell her about our trip to the crash site, the yellow tape, the bishop's prayer, the media's demand that we explain why planes fall from the sky, and the anxious eyes of the coroners as they waited to go to the site. But most of all, I wanted her to know that I was thinking about going and serving at the crash site. What would she think? Was this the Lord's call for me?

When it came time to say it, I couldn't. Bottom line: I still didn't want to go. I didn't even think that I could actually do it. My own track record had convinced me that I would get to the yellow tape and freeze—then turn, walk away, and not face it. So I told Erilynne nothing more than that we were going back to the Plaza to help however we could. It was best left unsaid. To talk about it would have made it more real, more possible. And I wasn't ready for that. I didn't have the confidence.

As the bishop left for downtown Pittsburgh, Ken, John, and I left for the Plaza. Each of us knew that the Lord was calling us to go to the crash site. But all the way, I wrestled inside. Old, painful memories came back, of a time when I once faced a similar decision and chose not to go. How I hated those memories. How I hated myself for shrinking back and not crossing the yellow tape of years ago. But that's what had happened.

And here I was, all over again.

Saturday Morning, October 27

When the knock came on my door, I was getting ready for a football game. I opened the door and found the headmaster of my Episcopal boarding school needing to talk to me.

I didn't want to be at the high school. I was a junior. Dad was living in San Francisco, soon moving to London, and traveling extensively in his job. That day he happened to be in Tokyo. There was no practical way I could live with him. My sister, Kate, was a sophomore in college. My brother, Gregor, was working in Pittsburgh. Mom, who was fighting cancer, had left Pittsburgh and was living with my grandmother in a suburb of Detroit some forty-five minutes from my boarding school.

Mom wanted me at the school. She knew that if she died I'd be in good hands. The headmaster was a friend of the family. He had kindly accepted me into the school. Plus, I was near my grandmother and other close relatives. It all made sense. Mom was pleased. I just didn't want to be there.

Still, in the face of it all, I was optimistic. I had made some new friends, joined the school band, taken a part in the fall theatrical production of *Antigone*, and worked hard on my studies. Where did this optimism come from? It was very simple: *Mom wasn't going to die.* She was going to beat the cancer. The doctor said that 2 percent survive what she had. That wasn't many. But I believed she'd be one of the 2 percent.

There had been a note on my pillow in late August, before I moved to Detroit. I had spent the day away from the house. I was sixteen and confused, not wanting to leave my home, and hating the fact that Mom was sick and I couldn't be there for her since I had to go to Detroit. It was all too much me for that day. When I came back in the late afternoon, Mom was waiting for me. She knew I was upset and wanted to talk it out. I don't remember our conversation now, but I still have the note she left for me later that night.

> I have felt so far away from you this day—that grieves me—for I love you so, and you are, by virtue of your nature, a very positive force in my life. When you came home, you told me I could beat anything, I think you can too!

I told her she could "beat anything"—and I believed it. I went to the boarding school convinced that I would be there only one semester. I dreamed of Christmas back in Pittsburgh, of our family reunited, of being with my friends, and then going back to my old school come January. First, Mom needed her chemotherapy treatments. That, plus prayer, would cure the cancer. She would be well by winter. We'd be home again.

To me, the cancer was never serious, never life-threatening. At one point, the cancer medication bloated her stomach. She teased on the phone, "It feels like I'm having twins!" We laughed. And why not? This would soon pass. Another time, I was offered a ride to go see her for a few hours. Incredibly, I chose not to go. I was having fun with my friends. I had school work. I was not aware that the time was short, that death was coming, and that I would soon not see her again. Nor did I realize that I would later look back on that invitation—at least a thousand times—and regret my decision not to go.

The headmaster knocked on my door that Saturday, October 27. He told me that my uncle Bruce had called. Mom had taken a turn for the worse. It was time for me to get to the hospital.

"Your sister, Kate, is already there and your brother is on the way," he said. He stayed with me as I put a few things in my suitcase. He had arranged a ride. I'd never seen him so tender, so solemn.

He knew she was dying. I saw it in his face. But I couldn't accept it. Even then I thought to myself, *She will rally! She will come back. You'll see, sir. You don't know my mom like I do. She's not dying!*

Saturday Afternoon, Five O'Clock

The hospital room was dark, the blinds drawn. I saw her face from the corridor but didn't recognize her. I thought I had the wrong room until I saw my grandmother on one side, my sister on the other. Kate came over to me, gave me a hug, and walked me to Mom's side.

Nan, my grandmother, was a regal looking woman with brilliant white hair, soft skin that never seemed to age, a sturdy build, and a strong yet gentle heart. Her unselfish character had withstood years of her husband's illness. She had buried him only five months before. Now, she stood by her beloved daughter, Joanie, rubbing her legs, holding her hand, and keeping her comfortable in her last hours. My dear Nan. She would live the next twenty years holding the grief of her daughter's death in her heart.

Kate had the same determination to be there for Mom, just as Mom had been there for her.

"She's changed so much," Kate whispered.

And she had. Her auburn hair was pulled back. It had thinned and grayed. Her face was jaundiced, old, and thin. For a moment, her once striking forty-five-year-old face looked like my grandfather's—as if she had aged thirty years. I picked up her right hand and held it. This was her hand—those distinctive, long and slender fingers. She held on, but I wasn't sure if she was aware.

"Talk to her," Kate said. "She knows you're here."

Mom's brown eyes went in and out of focus. Her breathing was heavy. Her head moved from side to side—looking, staring, turning. At one point, she looked right into my face, and her eyes focused. I drew closer. I told her I loved her. She kept looking. Then she was gone again.

I couldn't stay in the room. I could not keep watch with her.

I joined other family members in the waiting room. I wanted to be with her, but I couldn't. I didn't know what to do.

She wasn't going to rally. She was going to die. I had always been protected from death. It's part of life, but I didn't know it then. I wasn't prepared. Death wasn't supposed to come to our house. Our family believed that "all will be well." But that wasn't true. Mom was dying and I was scared to see it happen. Was she in pain? Would it hurt? Would she just close her eyes and stop breathing? What was going to happen to her?

There was a strong antiseptic smell on the cancer wing. I had never smelled it before, and I associated it with the smell of death itself. I took the elevator down to the main floor and found a phone. The overseas operator dialed Japan, then Tokyo, then the hotel.

"Malcolm Barnum, please," I said loudly to the receptionist over the static.

"Hello," Dad said. The sound of his voice made me cry.

"Dad, I'm calling from Bon Secours Hospital." I tried to compose myself. "Mom's dying." I described her condition. I told him family members were gathering and that I supposed it wouldn't be long.

"I'll catch the next plane," he said reassuringly. "I'll call you when I land in Seattle. I should see you some time tomorrow." He told me to keep praying, to hold my chin up, and to give his love to my brother, sister, and grandmother. I put the phone down, weeping.

Back up to the fifth floor. I stood in the corridor, leaning against the wall and staring at the entrance to her room from a distance. I wanted to go back in the room. But I just stood there—frozen. Through the years, I have pictured myself in that same corridor, wishing I could push back time, unfreeze my feet, and go to her. If only I could have been there while she was dying, to stand watch as her son, and to be present with her when she passed from this life into the arms of her Savior.

I didn't go. That was the yellow tape of my past, and my most painful memory was that I could not cross it. I could not face the moment of death with my mother.

At five o'clock, my sister came out of the room. Her hands were over her mouth, tears were running down her cheeks, and there was the sound of muffled crying. The family in the waiting room came out as Kate and I hugged each other. Uncle Bruce went into the room. A few minutes later, Nan and Bruce came out, closed the door, signed some papers, and joined the family.

"She didn't want me to be there," Kate said, half sobbing. "It's like she knew she was about to die. She signaled for me to leave. I knew what she was doing. Just as I started to go, she died."

"But you were there for her, Kate," I said, holding her, grieving Mom's death. I hated not being there to the last. I kept looking at the door to Mom's room, thinking I should still go in and pay my respects. *I must face her death with the same courage she faced death.* But how could I do it? I had never seen a dead body. The thought scared me. Still, I knew I should go. I asked Kate, "Shouldn't we go in and say good-bye?"

She wasn't going to, she said. But if I needed to, I should.

I hesitated. Nan said it was time to go home. We started putting our coats on. The elevator doors opened, the family got in, and I, still looking down the hall at Mom's door, knew I had a choice. "I'll be down in a few minutes," I should have said. It was time for me to grow up, time to throw away the child's view of life that "all will be well." Now I knew that that was a lie. Death is real. It eats the body, steals the mind, stops the heart, smells the halls, and stings those left behind with sorrow and deafening emptiness. I needed to walk down the hall and stay with my mother's body.

But once again, my feet froze. I couldn't do it.

Decontamination Suits

We pulled into the Plaza, parked, and got out of the car.

"Let's stay away from the media," Ken said, "or we'll never get to the site."

"I think we should go back to the Salvation Army trailer," John recommended. So we weaved our way through the congested parking lot, away from the media, and found Major Pfeiffer still inside the Salvation Army trailer working at his desk. We told him we wanted to go back to the site.

"It's now open," he said. "They sent the first bus of rescue workers about a half hour ago. I think it's great you want to go back. We need chaplains up there." He stood up. "Let me show you what you have to do. Come with me." He stepped out of the trailer, walked a few feet, and stopped.

"See that white truck with the back opened?" he said, pointing a few hundred yards away. "That whole area is roped off. The media and unauthorized personnel are not allowed in there. Now, do you see the group of people standing inside the rope? They're the next ones to go on the bus."

"Who do we talk to?" I asked.

"There's a large man with brown hair and a receding hairline. He's holding a clipboard and wearing a T-shirt that says 'HAZ-MAT' for 'hazardous materials.' There are HAZ-MAT officials here from every government level. In this situation, they're responsible for all areas of decontamination. Down here at the Plaza, they're outfitting every rescue worker in a decontamination suit. Up at the site, you'll see them again before coming back to the Plaza. Their job is to take your suit off and wash you down."

"We didn't wear decontamination suits this morning" I said.

"You probably should have," the major replied "You know, there's a health risk when dead bodies are exposed to the open air for any period of time. There's a strong smell up there already. In fact, to prevent the possibility of disease, they'll even decontaminate the plane before it leaves the site."

"So, you think this man from HAZ-MAT will let us go up?" Ken asked.

"I can't guarantee that. They're only taking a certain number of people. But give it a try. Tell them you're working for the Lord. How can they refuse that! And you guys come back and see us from time to time, okay? Tell us how you're doing. And if we can be of any help, we'll be right here."

We had made a new friend. Over the years, my image of the Salvation Army was of a person standing outside the grocery store, tirelessly ringing a bell at Christmas time. I had forgotten that their true mission was to serve in emergency disaster situations, and that they had a long-standing record of providing a Christian witness to the suffering and needy. My admiration for them grew measurably that day.

We made a beeline for the HAZ-MAT truck. The man Major Pfeiffer had described wasn't there, so we stood close to the truck but outside the rope. A few minutes later, the man appeared. The group of rescue workers already inside the rope gathered around him. He asked for their names, their professions, and who sent them. One by one, they called out as he wrote the information on his clipboard.

"Susan Parker, medical assistant, Allegheny General Hospital."

"Sam Richardson, also a medical assistant from Allegheny General Hospital."

I turned to Ken and John. "Let's just do it." They both agreed. We pushed the rope down, hopped over, and joined the insiders. We waited for the nearly twenty workers to finish giving their names. Finally, the man asked, "Is that it? Anyone else?"

"Thad Barnum, chaplain."

Still looking at his clipboard, he said, "Who sent you?" I remained silent. He looked up, saw my black clergy shirt and white collar and smiled. "Right, *chaplain.* Okay, I get it! I know who sent you!" A few people quietly laughed. "And this is Ken Ross," I added, "and John Rucyahana." There we were, the three of us, standing in front of him—extras with no official orders. The decision was up to him.

"Well," he said, pondering whether he should send us, "who

am I to stop you from going? God knows you're needed up there. As for the rest of you, it's going to be about a half hour. Then we'll suit you up, put you on the bus, and send you to the crash site. You'll be up there for about two hours."

He went back to his clipboard and said, "Barnum, Ross, Rucyahana—chaplains." We were in.

The House Built on Sand

Not long ago, I was at the wake of a friend of mine. As his body lay in the open casket, I couldn't help but stare at his beautiful five-year-old granddaughter. She stood on the kneeling bench beside him, holding his hand, patting him on the shoulder, touching the military metals on his uniform, and gently rubbing his cheek. The girl's tears came freely. Her heart was broken. She missed her Pap.

I admired her ability to see and touch death like that. I think she knew something that I didn't know until I was an adult. Death is real. It will come to all of us. None of us are exempt. All of us must be prepared.

In the years after Mom's death, I learned to stay with the dying. My frozen feet thawed, and I left the safety of the corridor behind and made my way into their rooms. Each step was a big step. One, over the threshold. Two, to plant my feet and not run from them or their family members, no matter what happened. I thought that the more I did it, the easier it would become. But it didn't get easier for me. I've never been able to protect my heart from feeling the anguish, both for the dying and for those standing watch. It always hurts to be part of somebody's pain. More times than not, it brings to the surface my grief of years ago when my mother died, and I, standing outside, was too afraid to see death come to someone I so dearly loved.

All of us must be prepared to face death. And in my life, that preparation was beginning.

A story from the Bible made a profound difference in my life. It showed me who I was and who I wanted to become. Jesus Christ spoke of two men who built their houses. The first man He called wise, for he built his house on a foundation of rock. This man, Jesus said, "hears these words of Mine, and acts upon them." But the second man was foolish, for he built his house on a foundation of sand. This man also "hears these words of Mine," but he made a choice. He did not act upon them.

To both men, "the rain descended, and the floods came, and the winds blew, and burst against that house." The wise man knew the storms were coming and was ready for them. His house "did not fall, for it had been founded on the rock." But the foolish man's house "fell, and great was its fall." He wasn't ready for the storms life brings. They caught him by surprise, and left him in ruins (Matt. 7:24-27).

I thought I was like the first man. I believed in God, went to church, and showed kindness to my neighbor. I thought that meant I would never suffer from life's storms. God would protect me. God would keep me safe since—isn't it true?—bad things happen to bad people, good things happen to good people! I thought I was safe.

And then the storms burst in on my house—cancer, boarding school, death, loneliness—and my life, my house, fell down. For a long time, I blamed the Lord for His lack of care, thinking it was all His fault. Soon I came to see that the Bible never promises, "Believe in God and be free from tragedy." It says the opposite. To believe in God is to understand the nature of this fallen world. The wise man knew about the storms, and he prepared himself. I didn't know, and I wasn't ready.

It was time to prepare myself.

First, this meant that I had to face the facts: Storms come to us all. None of us are exempt. Second, it was time to build my life on the foundation of rock. That meant I needed to hear the words of Jesus Christ and act on them. I needed to read the

Bible and hear it taught as God's word to my life. And I needed to apply those words in the strength of the Holy Spirit to my everyday life. That, the Lord Jesus said, was being like the wise man, ready for the storms, and ready to face the real world with suffering and death.

As I began to do that, my feet thawed out and I started taking giant steps. I learned to walk down hospital corridors, enter the rooms of the dying, and stay. Because of Jesus Christ in my life, the fears of the sixteen-year-old, who stood outside his mother's room, subsided. I was moving from a foundation of sand to solid rock.

But at the Green Garden Plaza, I felt sand beneath my feet again and those same fears. What was I doing inside the rope, waiting for the bus, preparing to cross the yellow tape? Had I gone mad? Who was I to think I could cross the yellow tape? Would I freeze, turn, walk away, and not face it?

Was I prepared?

Part II

Stepping Inside

Chapter Five

THE SPECTACLE

The forty-five-minute wait passed slowly. We talked with the rescue workers, mostly medical people from the various Pittsburgh hospitals. Their job was to work with the coroners. None of them felt comfortable about going to the crash site. Some were matter-of-fact: "You just do it. Somebody has to do it. Nobody wants to, but it's got to be done." Others were more revealing: "I'm hoping I can stomach it. It's my work; I have to go, but I dread going. It's bad enough working at a car crash, but 132 dead?"

Someone tapped me on the shoulder and said, "There's a man over there wanting to speak to you." I looked and saw a group of men, outside the rope about fifty feet away, waving for me to come. Some of them had cameras around their necks and notepads in their hands. One had a cheery smile on his face.

"I think we should stay away from the press completely," Ken said. "If we go over there now, it may look like we're going to the site just to attract the media."

"Exactly," John stated strongly. "People will misunderstand our motives. We must not give the impression that we seek personal attention. We are going for the Lord. Our witness must point to Him."

"I don't think it was wrong to give the interviews this morning," I stated.

"I agree," Ken said. "There's a right time to speak to reporters. People want to hear from clergy in times of crisis. And it's important that they hear the gospel and be comforted. But I think you do it and get out. We shouldn't be in the spotlight. And we don't want to feed the media's hunger to sell their stories. What sells is sensationalism, not the gospel. They want blood and guts."

"Hey, guys," another man said to us, "over there. Channel 11 News wants to speak to you."

"Look at that," I mused. "They're all around us—like vultures!"

"You guys are becoming famous!" one of the rescue workers laughed. "Why don't you just go and talk with them. Get it over with. The bus isn't here yet; you've got plenty of time."

"Not a chance," I started, half in jest. "I'm not going to the crash site to be on the nightly news. I'll go for the Lord, but no way I'd go for the press. I'm not that crazy!"

They were all around us, calling out, trying to get our attention. None of us expected it. We were too naive to know what's a story and what's not. Now we knew: Clergy inside the rope, standing with the rescue workers, waiting for the bus—that's news. We stubbornly refused to look at them. We kept our eyes to ourselves and focused on our conversation with each other and some of the rescue workers. Fortunately, none of the press broke through the rope. Still, it made the forty-five-minute wait seem like an eternity.

"I think we're attracting more attention by refusing them," I said.

"Yeah, maybe," Ken responded. "But our refusal isn't a story."

"We need to wait," John urged. "They know we're not coming. We have a work to do, and we're focused on doing it. We're telling them by our silence that we don't want to be their next story."

"I agree," I said.

The man from HAZ-MAT finally reappeared. His assistant jumped in the back of a large white truck that was full of supplies. The time had come.

"Okay, I'm going to read off your names," he began. "When you're called, go to the truck. You'll be given a decontamination suit, gloves, boots, and a mask. A couple of our guys have duct tape. You'll need to wrap the tape securely around the gloves and boots so they're attached to the decontamination suit. The bus is on its way down. When you're ready, get on board. We'll leave within the next ten minutes."

Barnum, Ross, and Rucyahana were the last on his list. A military man named Vic helped us get dressed. The one-piece decontamination suit was made of a lightweight plastic, much like the kind we put over our outdoor grill in the summer. It was light blue, and its airtight fit made it extremely hot over our clothes. The rain clouds had left in the early afternoon. The sun was out and beating hard on the suit. I was sweating in a matter of minutes.

I put the boots on, and Vic spun the duct tape around each calf. On went the gloves—the form-fitting surgical type. Then a second, baggy pair went over them. The duct tape secured the outer pair to my suit. A soft cotton mask for the mouth and nose hung around my neck by an elastic band.

"Now, when you're up there, you're going to shake people's hands and touch things. So, don't ever put your hands near your face," Vic warned me. It sounded so official coming from a Vietnam vet dressed in fatigues. "Don't rub your eyes, scratch your nose, or wipe your mouth. We don't want you guys hurting yourselves, okay?" He finished with Ken and John, went to the truck, and came back with a jar. He unscrewed the top and dipped his finger into the creme.

"This is Noxzema," he said. "It's a mentholated creme to cut the stench. You're going to smell decaying bodies up there, especially with the sun being so hot." Ken was first. Vic rubbed the creme at the base of each nostril. Then he put some in Ken's ears. "You're not going to believe me, guys, but the ears also pick up smells." Vic was all too sincere to be joking with us. The ears smell? I watched others put creme in their ears. Now, if I were a reporter, I'd think *this* was news!

"God bless you guys for serving up there," Vic said caringly when he finished.

We thanked him for his help and headed for the bus.

"Excuse me, sir. Can I get your name?" a man asked, holding a pencil and notepad. Why did he want our names? We kept walking to the bus, spelled our names, and got on.

"And what church are you from?" he shouted from outside the bus.

"Prince of Peace," I hollered back. "It's an Episcopal church in Hopewell." I turned, walked to the back of the bus, and sat down. "Can you believe it?" I said aloud. "Vultures, right to the end!"

A rescue worker sitting right in front of us turned back and said, "You guys, what a spectacle!"

The Press Release

The next morning our pictures—and our names—appeared in every major paper from Boston to Los Angeles. We had become news: Clergymen, dressed in decontamination suits, going to the site. The photo caught Vic putting Noxzema in Ken's ear. One caption read: "The Rev. Ken Ross, an Episcopal priest, has Noxzema cream rubbed on his nose and face—to combat stench—at the command center at the Green Garden Plaza yesterday as he prepares to leave to enter the crash site. Episcopal priest, the Rev. Thaddeus Barnum, is behind Ross." The photo missed John.

A similar photograph came out in the *New York Times, USA Today,* and the following week, in the September 19 edition of *Newsweek.*

We had become news and I didn't understand why. Was it unusual for clergy to serve in times of crisis? Haven't chaplains always served in war and major disasters? Isn't that our place? My curiosity led me to interview Rod Glass, a Christian television reporter, a few months later.

"Rod, what happened? Why did clergy going to the site make news?"

"We look for stories," he began. "Things that are unusual, that catch the eye. In the world's view of religion, God and clergy are separated from the real life people live."

"What do you mean?"

"Most people see God as distant, outside our world, and out of touch with reality. Let me give an example from my own life. Last Friday, I woke up to my kids screaming. I was late for work. The Cheerios box hit the floor, scattering cereal everywhere. The car didn't start right away. I got stuck in traffic, and by the time I got to my desk, I was exhausted. My 9:30 meeting bombed. My boss told me I'd have to stay late and redo the project I'd been working on. Meanwhile, late morning, I learned that my best friend had been served with divorce papers. He wants to have lunch and talk. I only have an hour. Back at the office, I get the news that my company is downsizing. My job's on the line—again. I finally get home. It's late. My wife has a fever, the youngest has strep throat, and the dishwasher broke. That's my day.

"So at last Sunday morning comes, right? We get ready for church, and nine times out of ten, we're not there on time. Getting three kids in the car, for whatever reason, never works. Anyway, we get there. Two go to Sunday school. The one getting over strep stays with us, moving, twitching, talking, making everyone look at us as if we're from a different planet. But what are we supposed to do? It's life.

"Okay, we finally get settled and guess how the preacher begins his sermon? He reads the opening verses of Psalm 23. Let me tell them to you, okay? I think it'll help you see my point." Rod picked up his Bible and began to read, "The Lord is My Shepherd, I shall not want. He makes me lie down in green pastures; He leads me beside quiet waters. He restores my soul; He guides me in the paths of righteousness for His name's sake." Then he looked at me and said, "Now, I'm just an average guy. How am I supposed to relate to that?"

"I don't understand."

"See, that's not my world. When do you think I have time to 'lie down in green pastures' and relax 'beside quiet waters'? It's impossible with three young children and a stressful job!"

"Yeah, but Rod, that's not what Psalm 23 is saying."

"I agree with you. But that's how the masses out there hear it. People think there are two worlds: God's and ours. His world is a big, gothic cathedral. Or better yet, it's like a monastery out in the country, where monks and nuns pray. They stroll the grounds in solitude. Life is peaceful and serene. That's the picture of green pastures and quiet waters. But that's not us. We live in the real world. It's the city and the suburbs—fast, loud, and busy. It's where kids throw up, bosses demand overtime, your best friend's life is in the pits over a divorce. It's where homeless people meet you on the sidewalk, begging for money. You think, *Where is God? Why is He cloistered in some cathedral or monastery, separated from the hell most people live in? Why isn't He here helping us out of the rat hole of everyday life?*"

"Most people think this way?" I asked. "Two worlds?"

"Yep. And the worlds are separated. God never comes to our world."

"But what, we can go to His? Just go to a church or monastery?"

"Well, yes, but think about it. When we do a story about a church on the news, we're usually in some dark, old, cavernous building. It's got hard, wooden pews. The few people there aren't even sitting together. Most of them look like they're eighty-five. The stained glass windows are from another century. The organ plays in the background, but who listens to organ music? The hymns are outdated. The priests and high altar are far away from the people. The clergy are dressed in robes. They almost always speak in low, monotonous voices reciting traditional prayers. This is God's world. Maybe this is what heaven is like, but it's not real life on earth. My generation can't relate. What happens in that kind of church doesn't have meaning in relationship to what happens in our lives. So God is seen as distant, out of touch, and separate from us.

"For example," Rod continued, "these kinds of services last about an hour. If it goes over, people get restless. Now, they don't get restless in the afternoon watching the Pittsburgh Steelers play football for three hours. Time flies by. Why? Because we live in a world of constant tension—offense, defense, pushing down the field, trying to succeed. We relate to football. It's part of how we think and live."

"But not church," I stated. "And it's probably the same with clergy? People can't relate to them because they're part of God's world—far away from the struggle and suffering of real life? Right?"

"Right. From the media's perspective, clergy are just as distant and out of touch. So when a pastor falls into scandalous sin, it's our lead story. It's a jolt for people. They don't expect it. Anytime the clergy step into our world, it's news. That goes for good stories, too. If a priest dressed up in blue jeans and a T-shirt and went into the ghetto to feed the poor, we'd be there. It's news. Stories like that sell. It's how Mother Teresa became famous. She's in our world, suffering with those who are dying."

"So that's it," I reasoned. "That's why three clergymen dressed in decontamination suits and going to a crash site made the newspapers. We stepped into the real world."

"It was news because it was the unexpected," Rod said. "You didn't belong there. You belong in church or a monastery—or in green pastures by quiet waters. You're supposed to be there praying for the rescue workers and the families of those who died. But you did more than that. You came into our world. And that's why the media was all over you. It's why the photograph made the national papers."

"What effect did it have?"

"A good effect," Rod answered quickly. "Those photographs gave witness to the Lord. They said that clergy go to crash sites. They get into our struggles. They relate to our lives. And that's important. As much as possible, I try to make that same witness in my work on television. I want people to know that God isn't separated from our lives and unapproachable. I want them to hear the next part of Psalm 23."

"Tell me what you mean."

"I told you my pastor preached on that psalm last Sunday. His sermon was like one of our news stories. He tried to shock us. He asked, 'Where do you think God keeps His green pastures and quiet waters? Are they close by or far away? Are they real to our lives?' Then he read verse four, which says, 'Even though I walk through the valley of the shadow of death, I fear no evil; For Thou art with me; Thy rod and Thy staff, they comfort me.'

"Our preacher said, 'The green pastures and quiet waters are found when we find the Lord. Even in our worst valleys, they are there!' He caught my attention! He told us never to fear 'the valley of the shadow of death.' The Bible promises that the Lord will always be with us when we call on Him. So you see, He's not out of touch with our real world. He doesn't peer over His heavenly balcony, see our suffering, and remain to Himself. He's right there—in the middle of our mess. I thank God for that. It's the only way my wife and I could handle three kids, stress at our jobs, and days when nothing seems to go right. I hold onto that promise. I know the Lord isn't far away. He's with me and my family."

"That verse," I told him, "stayed with me the whole week I was at the crash site. I felt as if I was actually standing in 'the valley of the shadow of death.' If the Lord had been nowhere near the crash site, I'd never have gone. What comfort could I have brought to the rescue workers if He wasn't there with us?

"For example, I remember one young man at the Plaza," I went on. "It was a few days after the plane crashed. He said, 'I'm not going back.' He was trembling, real nervous. Tears were running down his face. His voice broke when he said, 'They can fire me. Fine! But I can't do it. I can't eat. When I close my eyes at night, all I can see is death everywhere. What am I supposed to do with it—all this death?'

"It made all the difference for me to look him in the eyes and say, 'I know. I've been there. I've seen what you've seen. I know that pain. The images of death are with me, too. There's only one way I can make it. The Lord has promised to be with us in our deepest, darkest valleys. He said He would comfort us even in the midst of death. I have found Him faithful to that promise. So will you if you turn to Him.' "

"That's what it's all about," Rod said.

"In that young man's eyes, I knew how important it was for Christians—and for clergy—to go into the midst of suffering and death. We have to stand for the Lord, who does the same thing Himself."

"I agree. I want to do that in television. The world out there needs to know that God cares about our human tragedies. And honestly, I think that's the message your photograph gave in the newspapers."

I thanked Rod for his insight. I knew one thing for certain: The Lord was there the day I boarded the bus for the first time. The media may have thought we were a spectacle. Newspaper readers may have been surprised by our story. But we had to go. We knew the promise of the Bible: The Lord never stays outside the valley of the shadow of death. He's not far away. He's not out of touch with real life, not distant or separated from us. It's just the opposite. He meets us in the valley. That's His promise. He crosses over the yellow tape willingly, boldly, and confidently. And if He does it, who were we to stay behind?

On the Bus

We became quiet when the bus started to move. The media's hype, the Noxzema in the ears, and the conversations between us were distractions that quickly disappeared. We were on the way to the site. In a matter of minutes, we would get off the bus, head down the old logging road, and cross the yellow tape.

In my mind, I began to quote the words of Psalm 23: "Even though I walk through the valley of the shadow of death, I fear no evil; for Thou art with me; Thy rod and Thy staff, they comfort me." Over and over, I repeated the words, "I fear no evil; for Thou art with me." I needed Him. I still was afraid of getting to that tape and not being able to cross it. And if I did cross it, would I be able to handle the sights and smells of violent death? Would people see the chaplain unable to cope? What kind of an example would I be? How would I bring others comfort if my own soul was too deeply troubled?

"For Thou art with me; Thy rod and Thy staff, they comfort me."

If I had my way, He'd get me out of the shadow of death.

Rescue me. Save me. Promise me I'll never go back. But don't make me stay. Comfort me—by leading me out, not by keeping me here in the valley. Countless television preachers have become rich by promising that God doesn't keep us in the valley; instead, they say, He always takes us out. It's a message that sells. Who wants to hear that the Lord promises to stay in the valley of death with us, defending us with His rod, leading us with His staff? For many, that's not enough.

"Even though I walk through the valley of the shadow of death."

That's where I was going. As we turned onto the dirt road, faces came to my mind—people I've met who live day in and day out in that valley. They never leave. Their suffering is constant. The shadow of death is always there, always over them. They don't know life outside the yellow tape.

On a preaching mission to East Africa, Erilynne and I met a thirty-five-year-old nurse. She lives in the heart of the Third World. Every day, people line up outside her clinic in Kiboga, Uganda, suffering from fever, malaria, worms, dysentery, bleeding, AIDS, and pneumonia. Little children, malnourished, with bloated stomachs, look up at her with hungry, longing eyes. There's so little she can do. She only has certain medications. There is only one doctor, and he oversees too many clinics. He comes when he can.

This is her home. It's in the valley of the shadow of death. It will always be her home.

One young woman had a fever that wouldn't go away. Her husband, a banker, was out of town for a few weeks. She came to the Kiboga clinic with her two-year-old son in her arms. The nurse gave her medication, but it didn't work. She took the young woman and the child into her care for two weeks. She tried everything she knew. But the fever was too strong, and the young mother kept losing strength. The doctor finally came with stronger medication, but it was too late.

The nurse lives every day in that valley. Her heart breaks to see the suffering, the tears, the constant death. But she'd never leave. "The Lord Jesus Christ," she told us, "is my daily strength."

Another face, in Pittsburgh. A woman in her mid-seventies can't shake the aching loneliness. Her husband and dearest friend ("We met when we were six") died seven years ago. The pain, she told us, never stops. She's always sad, always on the verge of tears. She prays, goes to church, enjoys her friends and family. But she's haunted by the empty chair in the family room, the sound of his nervously tapping foot, the smell of his after-shave, the long hours of conversation, the endless memories, the other half of the bed—still made. She loved him so much. She knows she'll never leave the valley.

Another face, a waiter at a local restaurant who sometimes served us. I hadn't seen him in awhile and asked how he was doing. "I've got cancer," he said openly. "The treatments made me miss work."

"Are they over now?"

"Yeah, but the doctors told me it's everywhere. It started in the lungs. I had one lung removed last summer. We thought it was caught in time." He's thirty-six and has four children at home, all under the age of eight. His wife was diagnosed with cancer at the same time. Her breast was removed. The doctors think she'll make a full recovery. But he won't. It's terminal, a matter of time, and he knows it.

Other faces came back to me: A little baby boy, blind and deaf. An older woman crippled from arthritis, her pain nearly more than she can bear. A man in his fifties, suffering from a chronic mental illness, who has lived in a state mental institution for thirty years. I spent a summer on his ward as chaplain. We spent time together, and at the end of the summer, I left. He's still there. He'll never leave.

So many people walk through that valley every day—and never get out.

When I was a young pastor, I was an optimist. No problem

was too great—until I met people whose suffering was beyond repair. They were in the valley, and that was where they would stay. It was only then that I understood Psalm 23. There, in the shadow of death, the Lord has a promise. He meets us there. His presence with us means we don't need to fear evil. His rod and His staff bring comfort. There, in the darkness, He makes us lie down in green pastures. In the shadow, He leads us beside quiet waters. He restores our soul—That's His promise. And each day, it is enough.

The bus drove along the last stretch of new road. It was almost time for it to come to a stop, open its doors, and let me out. I could feel my heart racing. My hands were cold and sweaty. But my resolve grew more certain. I knew the Lord had called me. I knew He promised to be with me. I knew the Lord Jesus, who made the nurse in Kiboga strong to serve the sick and suffering, would do the same for me.

"I fear no evil; for Thou art with me."

Val

The bus stopped next to the Salvation Army canteen. As we got out, I saw a small group of men and women ready to board the bus for the Plaza. They had finished the first two-hour shift. Others were at the canteen, getting a drink of water in the hot sun. Just past the canteen, I saw a HAZ-MAT team at work.

Large, rectangular sheets of yellow canvas were spread out on the ground. On top of them were two small, plastic pools. One was filled with cleansing liquid. The other contained running water. To the left were trash cans for the used decontamination suits. To the right was a line of three or four rescue workers. I watched the first one, a man in his late forties, step into the first small pool.

He held his arms up at his sides. Two men from HAZ-MAT cut off his mask with scissors. Then they cut off the duct tape. Off came the gloves. They unzipped the front of the decontamination suit. They washed down his boots by dipping a brush into the liquid and rubbing vigorously. They ripped the suit off, then the boots. He stepped into the next pool of running water to clean his shoes, hands and face. It was all a matter of prevention. There were hazardous materials at the site, not to mention the risk of disease. The man stepped out of the second pool, and the next rescue worker came forward.

Two large trucks were parked to the left of the HAZ-MAT operation. The big bay doors in the back were open and facing the crash site. Two men dressed in dark blue jumpsuits were waiting inside one. We watched as two rescue workers walked along the path from the site, each holding one end of a blue container that was about a foot and a half deep, three feet long, and two feet wide. They hoisted it up to the men in the truck and went back to the site. Two other rescue workers carrying a bin were twenty feet behind them. And again, behind them, two more were coming.

Those trucks were the refrigeration trucks I had heard about. The dead were being carried out.

I looked down the old logging road for the yellow tape. It was gone. In its place was a dark green military supply tent. It was about thirty feet long, and it filled the road. From the Alpha side, no one could get to the site or leave it without passing through the tent. Beyond it, with the sun now shining on the crash site, I could see rescue workers everywhere—on the road, up in the woods to the right, and down in the ravine below. How different it looked from the rainy morning when only a few people were around.

We made our way to the tent. There were some older men standing outside it. One looked like a state policeman. They saw us coming and talked with us for awhile. We were the first chaplains there, and according to them, we were sorely needed. The more we talked, the more uncomfortable I felt. It was time to stop talking, walk through the tent, and go to the crash site. I didn't want anymore delays.

At the right time, we excused ourselves.

I opened the tent flap and walked in. The room was dark. My eyes were used to the sunlight. I saw a man to my left, sitting in a wooden chair, drinking coffee. A card table was set up next to him, with white plastic cups and a red, gallon-sized coffee jug. To my right were supply boxes stacked shoulder high. Ken, John, and I pressed ourselves against the boxes as rescue workers passed by carrying human remains to the refrigeration trucks.

At the end of the tent, talking with one of the workers, was Val.

I couldn't help overhearing her words. There was a brightness in her voice and a solemn yet gentle smile on her face. The man was having difficulties with his superior. Val knew the right way to handle it. She spelled it out, giving him suggestions. Then she patted him on the shoulder and told him to come back if it didn't work. He nodded his head, and with a gleam of hope in his eyes, he went back to the site.

She looked up, saw us, and greeted us as if we'd walked into her home. Val Tatalovich is a striking woman with a warm, energetic, encouraging presence. She and her husband, Wayne, the county coroner, were doing everything possible to make the best of the worst, and that meant making the tent feel like home. She took care of the supplies and made sure the workers had what they needed to do their jobs.

But Val had more to do, and she knew it.

Her heart was for the workers. When they came through the tent, she welcomed them. When it was needed, she comforted, gabbed, laughed, cried, cheered, and prayed for them. She knew many by name. And she knew the crash site and what it cost the soul to work there. So she set her mind to making the men and women feel appreciated and cared for in their service. She was a ray of light in a dark, gruesome valley—a messenger of encouragement with just the right words at the right time.

She stood next to the tent flap which opened to the crash site. I thought it odd, but so right that Val was in the exact spot where the yellow tape had been that morning. We talked for a few minutes. She said she was thankful we'd come. Chaplains were needed, and, she assured us, we had important work to do. Then she smiled, opened the flap, and sent us out.

Chapter Six

In Death's Shadow

We stepped onto the site. I felt awkward, with absolutely no confidence. Years ago when I was a chaplain-in-training at a hospital, I fought awkwardness. I would enter a room in which a patient was dying, where the family was gathered around the bed standing watch. In these situations, family members want to be there out of love for their own flesh and blood. For them, nothing more is needed. But for me, just being there, hanging around with nothing specific to do, felt awkward. I wanted to do something—anything—to help ease their suffering and calm my jitters. If I couldn't do something, I would start talking nervously. I'd use trite phrases. I'd bring up stupid subjects that were poorly timed, bothersome, and distracting to a grieving family.

Slowly, I learned a new confidence in facing a situation in which nothing can be done. I learned to stay, to be a quiet presence, to love without speaking, doing, or being noticed. It was hard for me, but I've known such love by experience in my times of grief. There's a physical comfort in having family and friends around who *want* to be there. They don't feel awkward. They're not looking for something to do or demanding to fix what can't be fixed. They're just there, present, at ease with themselves and with me. That quiet love can soothe a broken heart. Somehow, it can help us carry the heavy weight of grief.

One step onto the site of the plane crash, my confidence fled. I felt awkward again. I wanted something to do. Everyone at the site had a specific job to do—that is, except us. We were just there. Why were we there? *Should we stop the rescue workers and talk to them?* we wondered. *Do they want to talk? Is it the time and place to talk? Or do we find a place to stand and wait for them to come to us? But where do we stand? Where do we go?*

A mortifying thought came to me. What would people think if we roamed the crash site with nothing specific to do? There are people who are driven by morbid curiosity, who take a dark delight in seeing the sensational, the macabre. Is this how people would see us? My jitters increased, and my sense of purpose evaporated. "Lord Jesus," I quietly prayed, "help my awkwardness. Show me how You want us to serve here."

We were no more than ten feet past the tent when two rescue workers came toward us. It was near the spot where I'd first seen the body of a man in the morning. The body was now gone; a small flag marked the spot. The two men were carrying remains. I was standing just in front of John and Ken when I caught the eyes of one of the men and said quickly, "May God bless you for being here today."

They stopped.

"Thanks. That means a lot. Where are you from?"

"We pastor a church in Hopewell Township. We wanted to

be here at the site with you guys."

"I gotta tell you something," he said. "This brings back memories from Vietnam. I was there in '68, and I can't tell the difference between the aftermath of a plane crash and an air raid. Bombs do the same thing."

"I was watching TV with my kids," the other man said. "My wife heard the explosion from the kitchen. She knew something awful had happened. We went out the front door and could see the smoke coming up. I told her, 'That's a plane crash.' I knew it, but I didn't think I'd be involved. I went to work this morning and my boss said, 'You go help at the crash site.' I called my wife. She's really upset that I'm up here. She's afraid for me. Would you guys remember my wife in your prayers? It would mean a lot."

They put down the bin holding the remains. They wanted to tell their stories—things like where they were when the plane crashed, why they were at the site, what it was like to see the site for the first time. They both gave graphic descriptions of the mangled bodies they'd seen. They needed to say it, to get it out. It was painful for them, but there seemed to be some relief in talking about it.

"Are you Christian men?" I asked.

"I haven't gone to church in years," the second man said. "I know I should, but today, I've never prayed so much in my entire life. I thank God my mom and dad brought me up to know Him. I always used to read my Bible. My wife and the kids go every Sunday. I'm gonna go this week. How can you stand here and—," his voice cracked. He looked away. He bent down and started to pick up the bin.

"We'll be praying for you and your wife—for both of you," Ken said, looking at the two men.

The first man bent down and picked up the other side of the bin. "We need it," he said matter-of-factly. His eyes were watering, the emotion too close to the surface. It was best to keep going, put the next foot forward, push the emotions down and not think too much. There was work to be done. The old, bitter memories of Vietnam, as well as the terrifying sights of today, would have to be dealt with later.

"Gentlemen, remember something," I said. "You're doing important work here. You're serving the families of the dead. They can't be here to see this mess. But you're here in their place, honoring them, their dead, and the Lord who sent you here. I know it's hard, but I'm thankful for your service."

"I know," the first man said. He nodded, looked at me kindly, and turned toward the tent.

We took only a couple of steps up the logging road toward the impact site.

"You guys are chaplains?" a man in his forties asked. He was working by the side of the road.

"Yes, we're pastors of a local church," I responded. With that, he began to tell his story. No one was more surprised than I was. Some of these rescue workers wanted to talk. Just yesterday none of us, including the rescue teams, imagined we'd be here on this road, facing terror and death at this magnitude. For whatever reason, it felt good to talk and let the story out. The words came from deep within—full of emotion, passion, grief, and fear—sparked by a crazy rush of adrenaline as we stood among so many dead bodies. These men and women described their last twenty-four hours in vivid detail, step by step, up to that very minute.

My prayer had been heard; the awkwardness was leaving. I now had a reason for being at the site.

Dan Drotar's Story

Dan and his wife had finished dinner, driven to the local mall, and were taking their evening walk. Their lives were in transition. They had hoped the timing would be better, but they sold their house before their new one opened up. They were staying in a hotel for a few weeks. As a result, their routines were different. Nothing felt like home. They were trying to make the best of it.

At a brisk pace, they passed the entrance to Sears. Bang! Bang! They heard two loud explosions back-to-back in a split second's time. It must have been, Dan reasoned, a jet engine backfiring. They were close to the airport. Dan is a USAir worker. The worst never dawned on him. They kept walking, wondering what had happened. In a matter of minutes, fire alarms sounded, horns blew, and high piercing sirens wailed from every direction. Fire engines, ambulances, and police cars raced past the mall.

"We'd never heard anything like it. We thought there must have been a terrible car accident."

After their walk, they couldn't drive home on Route 60. Both directions, the one to the airport and the other to downtown Pittsburgh, were closed. They had to find another road back to the hotel. Once there, they flipped on the television to a movie channel and sat down on the sofa with hot cups of tea.

The phone rang.

"Dan," his USAir coworker said quickly, "are you watching the news?"

"No. What's going on?" He turned the channel to a local station. A reporter was interviewing a state policeman at the Green Garden Plaza. In bold letters across the bottom of the screen were the words, "The Crash of USAir Flight 427."

"Oh, my God!" Dan sat at the edge of the sofa in shock.

"I called the office," his coworker said. "They're gonna need people to help. Nothing specific yet, but they want us to be there tomorrow morning at seven o'clock."

"I'll be there." Dan put down the receiver down. The double explosions, the sirens, the highway closed off—it all made sense. That night as they watched the news coverage, Dan's conviction grew deeper in his heart. He knew he had to help. He couldn't wake up, go to his desk, and work a normal day. He had to offer himself. He had to serve his company no matter what.

Morning came after a restless night's sleep. He was at the office by seven o'clock.

The work piled on his desk could wait, the boss said. He was needed in the "situation room," where tasks were being assigned. "We need you on the baggage collection team," they told him and six others. "The coroners will get the bodies out as quickly as possible. Then you'll go in and start collecting the debris. There'll be a USAir official at the crash site to give you further instructions."

A little after half past seven, the seven-member team was on the way to the Plaza. When they arrived, a USAir representative told them there had been a delay. The baggage collection team wouldn't get to the crash site until late morning. They'd have to wait a couple of hours. The time dragged, but eventually an emergency vehicle came, looking more like a big pickup truck loaded with equipment. They hopped in the back, drove up the hill, and entered the Beta side. A USAir official met them. He stood by a half dozen pieces of dark blue luggage with the USAir insignia on both sides.

After a brief introduction to the site, he said, "This luggage is filled with USAir emergency supplies. You should find rain gear and gloves. Take whatever you need. I'll be back to give you your assignments in a few minutes." He turned, walked back toward the site, and disappeared from view.

Dan Drotar looked down the old logging road. From where he stood, he knew that he was close to the site of impact, where the plane had actually crashed into the hillside. He broke away from the six other people and walked slowly toward the site. Fifty feet out, he stopped. His eyes focused on human remains lying in the middle of the road some thirty feet in front of him. They were not recognizable. They had no form. It didn't seem real to him. He kept staring, trying to take it in.

To the right, he saw what looked like one of the wings of the plane. As the rain came gently down, he saw steam still coming up from the wreckage, as if the fire wasn't completely out. A strange smell filled his nostrils. He thought about it, and then he knew. Jet fuel. He stood there quietly, looking and listening. There was an eerie silence. No birds were singing, no squirrels running through the woods, no mosquitos buzzing overhead. Except for the light rain falling on the leaves, everything was lifeless. A chill went through him. He shut his eyes and bowed his head.

"We made a wrong call," the USAir official said as he walked toward Dan.

"I'm sorry?"

"There isn't going to be a baggage collection team." The two of them walked back toward the other six. "If you want, you can go back to work. It'll be three of four days before we're going to need you. The county coroner said there's no way we'll get these bodies out until then. It's going to take time. They're scattered everywhere. So check with the 'situation room' on Monday. They'll know our progress and when we need you. All I can ask is that you do your best to come back."

"What's going to happen today?" one of them asked.

"There'll be coroner teams going out, collecting bodies. That's the first priority. Look, I've called down to the Plaza, and the pickup truck is on the way back. I want to thank you for offering your services. Your willingness and courage are exemplary. I hope to see each of you here early next week." The matter seemed to be closed. The USAir official turned back down the logging road toward the site.

"I'm going to do it," Dan confided to his coworker.

"Do what?" she asked.

"I'm going to collect bodies. I have to."

"You're kidding."

"No. I just can't go back to work and sit at my desk. I've got to help—if they'll have me."

The Drive to Serve

"Excuse me!" Dan called out. He had made his decision. The USAir official was still within earshot. He heard the cry, turned around, and saw Dan coming toward him.

"Sir," he said as he approached the official, "I have to help. I want to be on a coroner's team."

"Do you think you can do that?"

"I'm not sure, but there's one way to find out."

The man smiled, as if he liked Dan's response. "All right, you're on the team."

Just like that, his response was immediate.

"Go get your stuff—and hurry. The county coroner is setting up teams right now."

Dan thanked him and headed back. As he was walking, he wondered what he had just done. He knew he'd never be able to handle this job unless he turned to the Lord. He needed Him. It was the first time since the plane had crashed that he felt frightened. The realization hit him: He was going to work with the dead. He'd never seen death like this before. He wasn't a medical man. He worked for an airline in customer service. He wasn't prepared to see mangled bodies. Sure, he'd watched some movies, but this was different. It was real life. *Real death.* The thought scared him.

"Lord," he prayed, "I need You. I know I'm supposed to be here. I can't go back to the office. I have to do it. Help me. Be with me. Take my fear away. Let me serve You. Let me serve the families. Let me serve their dead. I can't do this on my own. Oh, God, help me, I pray. Help everyone who works here. And I ask this in the name of the Father, the Son, and the Holy Spirit. Amen."

He no more than finished his prayer when he saw the pickup truck arrive. As he rejoined the other six, he wasn't surprised to hear that his coworker was leaving. She looked at him with amazement and said, "I can't do that." Dan understood. He told her he wasn't sure he could either. But he had to try. Those words stayed in her heart and became an inspiration to her. Later that afternoon, she went back to the Plaza, signed up, got a decontamination suit, rode the bus, and joined a coroner's team.

After the pickup left, Dan got his supplies and followed the USAir official down the logging road. Within seconds, he entered the crash site for the first time.

Some months later, I interviewed Dan.

"Why did you have to go?" I asked him. "You keep saying that you knew you were supposed to be there. Why? How did you know that? What was driving you?"

"Couple of things. First, I had a deep feeling inside. I had to do it. When the USAir official told us to go back to the Plaza, I couldn't go. I knew in my heart that I had to stay and work."

"So it was instinctive. You just knew it. That was your main reason for working there."

"Partly. But I'm in customer service. It's a perfect job for me. I like to help people. In this case, all I could do was think of the family members. Their dead loved ones were at the crash site, and they would never see them again. I wanted to help them. I thought to myself, 'If I were a family member, I'd want a caring person to be there to retrieve the one I loved. I'd want somebody who would treat them with dignity, respect, and honor—somebody who would go the extra mile.' And that was me. I knew it.

"Of course," he went on, "by noontime, I had heard rumors that said only 20 to 25 percent of the victims would be identified—maximum. I heard those same figures on the nightly news. To me, that was an incentive to work all the more. In some ways, I felt helpless at the crash site since there were no survivors. But also, I felt a challenge. I wanted to work as hard as I could. I wanted everyone to be identified. I wanted each family to have something from their loved one—a ring, glasses, a wallet, a watch—something. The whole week I was there, I kept thinking about the family members."

"That's a noble and godly drive," I said, inspired by his words.

"Well, the Lord had a lot to do with it," he said.

"What do you mean?"

"For example, I worked at the site the whole week. Some people were on shifts. They stayed two hours one day, two hours the next. The Lord gave me the strength to be there every day, all day. I never felt I had to leave! Not once. I had the ability to stay and work, and I knew where it came from.

"But there's more," Dan went on. "I'd never be working at USAir if it weren't for the Lord."

"How so?"

"I had applied a couple of times, but it didn't seem to work out. At one point, I felt that my career was at a standstill. I wasn't happy, and I didn't know what to do. I sent out my resumé, made calls, and went to interviews. Zero. It was as if I hit a brick wall everywhere I turned. Depression started setting in, and I couldn't seem to shake it. My life was going nowhere fast.

"Now, you have to know I'm a Christian," he continued. "I accepted Jesus Christ as my Savior some time ago. I know He died for my sins and saved me. Well, one night I was driving down the road. I'd never felt so empty in my entire life. I cried out to the Lord for help. I was lost inside. I didn't have purpose for my life, and I was afraid. The depression was getting worse. Something had to change.

"Right there in the car," Dan said as tears filled his eyes, "the Lord spoke to me. I heard Him in my heart, as clear as I've ever known. I'll never forget it." He paused. The moment was right there for him, as if it had just happened. The words were still ringing in his ears. He closed his eyes and dropped his head in his hands. "The Lord said, 'Be patient. I'm going to take care of you.' With those words, I felt this peace come over me. Three weeks later, I got the job at USAir. I've always known that He gave me that job."

All the pieces were in place, knit together in perfect order.

"And because of it," I said, "you were there to serve at the crash site."

"I know it. I thank Him for sending me—for giving me the desire to go, for letting me care for the families, and for letting me help others. It's what I like to do the best."

Working at the Site

"So what happened after you entered the crash site?" I asked.

"I followed the USAir official down the road and through the site," Dan remembered. "It was my first look at the destruction. I saw body parts everywhere. Some, I recognized. Others, I didn't. I can tell you, I felt sorry for any rescue worker who was forced to go to the crash site. Maybe their job required it, or their boss demanded that they go. In my opinion, they should've been given a choice."

"Why?" I asked him.

"It's different when you don't want to be there. I saw some guys—that had a job to do and they did it, but down deep, the job wasn't driving them. They were afraid for themselves—afraid to see the dead, afraid the memory would haunt them for the rest of their lives, or afraid they'd break down and never be the same. Some of them just left—walked away. I felt badly for them. They never volunteered, never chose for themselves. I think some people will suffer for a long time because of it."

"That wasn't you," I stated factually.

"I chose to be there. I knew why I was there. I mean, I didn't eat for a couple of days, and it was hard to sleep at first. But I got up every morning and wanted to go back to the site. My goal was that 100 percent of the people be identified. Those families deserved it, and I'm real proud of our work. Hundreds of people devoted themselves to the county coroner and his staff. I had never seen anything like it.

"I mean, I heard some incredible stories," Dan said in amazement. "There were people from the medical profession, funeral directors, all kinds of people who closed down their work for a whole week. Many of them had no pay. They volunteered their time—some at the site, some at the morgue. They wanted to be there, to help in any way. They were just as driven to identify the 132 victims as I was."

"And it worked," I said.

"Yes, it did. When it was all over, 125 were identified. That's almost 95 percent. It made us feel bad for the families of the seven victims who weren't identified. But compared to the first estimates of 20 to 25 percent, it's a miracle. I still can't believe it. The Lord was so gracious to us."

"Yes, He was—and to the families," I added. "Dan, I don't understand something. Are you saying that it was easier to see the dead bodies because you were on a mission to help the families and identify as many bodies as possible? Did that make it less personal? Were you less affected?"

"Well, let me tell you what happened. I met with the county coroner and was assigned to a team of four. We were given an area of the crash site to cover. On our team, we had a pathologist and a coroner. Their job was to identify the human remains. They'd bend down over the body and examine it carefully. When they were done, the pathologist would describe the remains in detail.

"My job," Dan said proudly, "was to write down the description, then the location. The site was marked out like a grid, so every area had a location number. Also, I looked for physical items near the body. For example, at one point I saw a man's wallet a few inches away from a body. It was still intact and had his driver's license, credit cards, and some photos. I noted that on the paper. We thought it might help identify the remains. Actually, that happened a lot."

"What would you do next?"

"There was a state policeman on our team. He was the

photographer. The pathologist and the coroner helped him get pictures from different angles. Then they would place the body into a transparent bag. We'd slip my description in and leave the bag there. It was someone else's job to pick up the remains and take them to the refrigeration truck. Also, we put a fluorescent orange flag next to the body. Those small flags stayed there the whole week, marking where the dead had been found.

"Now, I haven't answered your question," Dan noted. "It's not that I didn't take it personally or that I wasn't affected. It's the opposite. I've never felt grief like that. But knowing I was helping the families put it in perspective for me. I remember the first time our team took a twenty minute break. They went up to the Salvation Army canteen for a drink. I joined them after a while, but first I had to walk through the site. I had to see it. I had to know what I was dealing with in order to continue."

"That helped you cope?" I asked.

"Yeah. It's like I wanted to get the worst over with so I could focus on my work. It was hard for me to stop the adrenaline. I was pumped, like I was suddenly a workaholic. I didn't want to take a break. I wanted to keep going, and I knew that wasn't right. So I thought I'd settle down more if I got used to my surroundings. And that happened, somewhat. As the afternoon went on, I felt more relaxed with myself, with the other team members, and with seeing death all around me.

"Of course," Dan revealed, "there's no way to distance yourself. Not really. We were dealing with people just like ourselves. That brings it home—that it could just as easily have been me. There was the pain of seeing the dead, then to think how they died, so suddenly and violently—and to know that you're dealing with men and women who were alive and healthy one day before, and with no idea this was going to happen. Anyway, even though I kept focused on my work, I couldn't stop the feelings of grief."

"I know," I said, remembering so well, "like the first time you and I saw each other."

Our First Meeting

My first two hours on the site went by quickly. That first day, John, Ken, and I stuck together. We made our way up the road slowly, visiting with rescue workers who needed to talk. Some wanted to pray. But there were also moments when we were left alone—left to see this world firsthand.

It was hard to take it all in. There was so much chaos. Ken would see something a few feet ahead and describe it to me. I'd look, but I wouldn't see it. There were too many things on the ground, in the bushes, and in the trees. Many of them were familiar: magazines, clothes, shoes, pencils, pens, soda cans, address books, appointment calendars—things I recognized. But there were too many things I had never seen before: the insides of a plane—or of the human body. It was all there together. I'd look. I'd look more carefully. Finally, I'd see what Ken was seeing. But it wasn't easy. It was hard to focus—and hard to want to focus.

For some reason, I wanted more order. I had thought that the plane parts would be bigger and more distinct, more obvious. Where were the seats that had filled the inside of the plane? Where were the 132 pieces of luggage? Where was the nose of the plane and the silver, cylinder-shaped body? Where were both wings?

All were gone; all were just part of the wreckage scattered on the ground.

In the moments we were left alone, we stood by the coroner teams as they worked with the dead. We saw the process as the pathologist and coroner examined the remains, described them to the note-taker, and helped the photographer take pictures. We saw the remains bagged—gently and with dignity—the flag put down, and the careful transportation of the body to the refrigeration trucks.

We were quiet as each body passed, in honor of the dead. Often, one of us would pray, "Lord Jesus Christ, be with this person's family right now. Comfort them in their loss. Ease their grieving hearts."

At five o'clock, the county coroner's staff called it a day. We had fifteen minutes to wrap up our work, head down to the tent, and take the bus back to the Plaza. The site would open at nine o'clock sharp the next morning. When the announcement came, we were standing on the logging road just below the impact site. We could see it from where we were. The hill went up steeply twenty feet, then flattened out. Another fifteen feet from there was the place where the nose of the 737-300 hit the ground.

We didn't make it there the first day. It was time to call it quits. Just as we turned to head back down the road, we heard a coroner say, "Hey, what's this?" John, Ken, and I looked at him.

The coroner squatted down. The pathologist joined him. They brushed the burnt ashes away and removed the debris surrounding it. Two state police photographers drew near and bent over to look. Another man, the note-taker, stood by watching studiously. The coroner and pathologist manipulated the remains, carefully separating the dirt from the flesh. They lifted it up, slowly, gently. For a few minutes, they worked as if it was routine. They had spent the afternoon doing this kind of work.

But then they stopped. They both froze. I could see tears welling up in the coroner's eyes as he held the remains with both hands. The photographers stood up, shaking their heads. The note-taker began to cry quietly. He looked at me and I at him.

"It was the first time I really lost it," Dan later told me.

I stooped over and looked again at the remains. The face. The hair. And then I knew.

"This is a child," the coroner whispered.

There was silence. Tears. None of us knew him. None of us had to. We wept for his death.

The Day's End

Part of me wanted to run from the site and never come back. But I couldn't. I had become like the family who must stay at the bedside of the one they love. No one wants to be near suffering, but there are times when we must. We stay with the dying because we love them. It costs us. It breaks our hearts. It drains the body. It confuses our senses. It makes the soul empty, lonely, and scared as if nothing in life makes sense. But the one thing it doesn't do is touch the will.

Family stays with family through it all. It's how love works.

Dan woke up the next morning and headed back to the site. He had to be there. He was driven by his love for the victims' families. They needed someone to care for their dead. The families would never know Dan, his love for them, or what it would cost his soul to work there. But that didn't matter. Jesus Christ had sent him. He was with him. And that's all Dan needed in return.

That's why we were there. The Lord had poured that same love into our hearts for the rescue workers. It's what brought us back the next day. We had to be with them when they worked with the dead—or when they wanted to tell their stories, pray, cry, or just stand there in silence.

It's hard to face suffering. None of us want to do it, but there are times when we must. And in those times, the Lord is faithful. He gives us strength. He meets us there. And we find the ability inside our hearts to stay with those we love, to come back the next day, and to not run from pain.

It's how love works.

Chapter Seven

WHAT'S IMPORTANT?

I ended my interview with Dan by asking, "What affected you most about working at the site?"

He didn't hesitate. "Sudden death. Until then, I'd never given death a second thought. I've had people close to me get sick and die. I've grieved death, and I know that one day I'm going to die. But before the crash, I kept pushing it away from me. I'm forty, so I figure I've got years in front of me. It's not time to think about death. When the time comes, then I'll deal with it."

"That's how most of us face death," I added.

"But seeing sudden death changed all that."

"How so?"

"I realized it could happen to me. When I saw those dead

bodies, I saw myself. I could just as easily have been on that plane. I fly all the time, business and pleasure. I'll bet I flew on that specific plane a half dozen times. Every day at the crash site, I saw myself aboard Flight 427. I kept thinking about the last minute of their lives. It was only a one hour commuter flight from Chicago to Pittsburgh. It was a beautiful day—a smooth trip. I wondered if some of them were resting. That's what I do when I'm flying in at the end of the day. My eyes are closed, and I'm entirely confident that I'm going home. My future is secure—no worries and no idea that this might be the last minute of my life. Then it happened!" He snapped his fingers sharply.

"Just like that—in twenty-three seconds!" He paused, his eyes wide open. "It's like I woke up to the fact that death can come to me just like that. It doesn't have to announce itself. For the most part, my life was like Flight 427—a smooth trip on a beautiful day. Never did I think I could be flying somewhere, driving in my car, sleeping in my bed, or even just sitting at my desk—and that's it! Death can be that sudden."

"That realization changed you."

"Before, it's like I was living in some pretend world. I thought, 'Don't think negatively. Of course, I'll be here tomorrow. Where else would I be?' I lived life as if death couldn't touch me. I was immune. Why make out a will? Why go to the doctor's office for an annual checkup? 'I'm fine. I've got a great pension building up for my retirement'—that was my attitude. But after the crash, I couldn't think that way. I knew I'd never be the same. I had to face life knowing I might not be here tomorrow. And honestly, sometimes I don't like thinking about that. It's hard to face the real world."

"It is," I admitted. "We want to think we're different—that we're exempt from tragedy."

"There's an emptiness about it. It hit me at the site, seeing all their personal effects. They were strewn all over the ground, all their belongings—trashed. Pieces of a laptop computer. Parts of a brief case. Business letters. Handwritten notes. A checkbook. Photographs in a wallet. A woman's makeup kit with a broken mirror. A watch stopped at 7:03—the exact moment they lost their lives. It felt as if I was walking into the home of someone who'd just died. All their possessions were left behind—things that meant something to them, things they valued. What good were they now? What value did they have?

"What happened to their goals and ambitions?" he went on. "What about their dreams? Had they made their mark yet? Had they finished their work? Were they ready to leave their families? Were the kids grown? What about the grandkids? Was it time to say 'good-bye' to their husbands and wives? Had life been full for them? Had they climbed all their mountains—walked all their beaches? Were they done living life? But how could they be? The average age on that plane—from what I heard—was forty years old!"

"I think that's about right," I said. "And no, how could they be done living?"

"That's the emptiness I felt."

"So you were grieving their deaths."

"Theirs—*and mine,*" he said emphatically. "Ten days after the crash, I was back at my desk as if it had never happened. My regular routine had kicked into gear. All I could feel was this emptiness. It's like my whole world was upside down. I kept thinking, 'What's important? Am I making a difference?' I used to finish a day's work and feel so satisfied, like I had accomplished something. Now, it was the exact opposite. I was restless both at work and at home. Nothing was the same. My house, cars, all the stuff I'd worked so hard for didn't mean much anymore. I couldn't shake the emptiness."

"What did you do?"

"My wife and I talked a lot. She's a trained counselor. We slowly worked through the grief. But in the end, I had to get a new perspective on life. I couldn't go back to a pretend world and make believe that the crash didn't happen—or that I'm not going to die soon, or that sudden death won't come to me.

"We prayed together," he continued. "A passage from the Bible was a big help to me. Jesus Christ told a story of a rich farmer whose land was very productive. He built big barns to store all his crops. Finally, he said to himself, 'Soul, you have many goods laid up for many years to come; take your ease, eat, drink, and be merry.' And that's what he did. Actually, it's what I did. But God came to him and said, 'You fool! This very night your soul is required of you; and now, who will own what you have prepared?'

"I saw myself in that position," Dan admitted. "That man lived life without thinking about death. His money became everything. It guaranteed him a long future and an easy life. He didn't realize that in one night, in a split second, he'd be dead. His soul was required by God. He had to stand before the Lord and give an account of his life. Was he ready? Was God first in his life? Did he ever think about Him? Then, bang! He dies a sudden death. Meanwhile, someone else gets all his goods. And Jesus ended the story by saying, 'So is the man who lays up treasure for himself, and is not rich toward God' (Luke 12:16-21).

"That's the perspective you were looking for!" I said.

"It changed everything. I don't want to live like that man. My possessions, my work, and my money—they can't promise me anything. I want to live for the Lord. He has given me a certain number of days on this earth. I want to know Him and serve Him. I want to live every day more fully. He's given me the desire to help people. That's what I want to do—in His strength. I want to be 'rich toward God.' "

"Sounds like you're prepared to die."

"I don't want the Lord to call me a 'fool.' I want to be ready for the day He calls me home."

"You're reminding me of a saying my wife memorized growing up in church," I told him. "It goes like this: 'Only one life 'twill soon be passed. Only what's done for Christ will last.' "

"That's what's important," he said as he smiled. "That's what I want in my life."

Running on Empty

Just after Mom's funeral, Dad and my sister, Kate, took me back to my high school dormitory. We walked through the campus, met some of my friends, went to my room, and talked. And then it came time for them to leave. Dad was flying back to California. Kate was off to college. We'd meet again in six weeks at my grandmother's for Christmas. With tears, we hugged. I watched them get in the car and drive away.

I had never been so alone.

In my pretend world, I had planned on being home in Pittsburgh for Christmas—with Mom and my whole family. We'd have a traditional Christmas, just like always. Then I'd go back to my old school, with the friends I'd known since fourth grade. That hope was how I had survived the first two months at boarding school. Now that hope, like everything else, was dead.

My small room became my home—just a single bed, wooden desk with a hard, wood chair, a soft, green chair with a reading light (Mom had given me that; she had wanted me to have her passion for reading, which I have), a little table for my record player, and a closet. It was all crammed close together in a tiny space.

The emptiness that drove Dan to face death was driving me then. But I couldn't face death. When my pretend world collapsed, I simply built another. I put my energies into school work. I filled up my time with sports, drama, and music. When the day was done or if I felt lonely, I'd slip into my room, shut my door, play my records, and read my books. They were all I had. This was the only place I could call home.

I didn't escape with drugs. I didn't drink, party, or sail in and out of relationships. But my escape was as profound. I was running just as hard. I sat in my chair, alone, not wanting anyone else to own my heart. If I loved again, I would hurt again. It was that simple, and I didn't want to take the risk. So I went to my safe little world, listened to my albums, read, and made believe everything was all right.

Death had come, and I didn't want to deal with it. Deep inside, I knew emptiness, and it hurt. I also knew that my one-man pretend world wasn't working. It didn't satisfy the fervent questions of my heart. I needed answers: Where was my family? Why do we love when there's no promise of tomorrow? Is it foolish to love fully, to give oneself sacrificially, and to pledge undying devotion when, *pow,* death can come at any moment? What can be trusted? How could I fall in love, marry, and have kids? Would the same thing happen again? Would they find *me* dead—gone one morning having left them behind? Or would they leave me, suddenly and unannounced? Would I have to walk through grief again, experience its emptiness, its loneliness? Would I be left with memories of the past, no hope for the future, face to face once again with the real world where death is, as always, a part of this life?

I kept moving. I worked summers, graduated from high school, and went to college at the University of Michigan. My heart was still set on going to seminary after college. I didn't question this decision. Instead, I kept my nose in the books, took full course loads, and took even more classes in the summers. I filled my emptiness with a busy schedule, and when there was a lull, I helped out at my church or played racquetball with my friends. There was no time, no desire to think about the past or to ask my questions. And I knew the paradox: I was going to seminary. I wanted to be a priest. But I couldn't turn to the Lord and ask Him my questions because deep down the questions scared me. How could I ask, *Why did Mom die? Why didn't You heal her?* It would make Him responsible.

What's Important?

It took me only three years to go through college. I was running fast and hard, but running on empty.

In early September of my last year of college, I had dinner with my priest and his wife. He knew about sudden death. He had been married three years, was just out of seminary and pastoring his first church, when his wife and four-month-old baby daughter were killed in a car accident. He remarried years later, had children, and now had grandchildren. He had a gentle manner about him; he was wise and understanding.

We were having a benign conversation, just catching up on each others' lives. I wasn't ready for what was about to happen. Like a doctor, he prodded me with questions, found the sore, and poked.

"I thought this was your third year," he said, puzzled.

"It is. But I've taken enough courses to graduate this year. I hope to go to seminary next fall."

"Sounds like a marathon."

"Not really. It hasn't been that bad. I've worked hard—gone through the summers, you know."

"So you're running and never stopping. All this because you want to go to seminary?"

"Yeah, well, I've wanted to be a priest since I was seven. So why not go for it?"

"I bet you're running for some other reason." He said it slowly, tenderly, as if he'd once been in my shoes and felt the push to run—as if he saw not the goal in front of me, but the fire behind me.

"I don't know what you mean."

Then he told me the story of his first years pastoring a church. He told he of his beautiful new daughter and the love he felt for his wife—a high school sweetheart. He described the dreams they had and his confidence in love, in life, and in the Lord who made the sun shine, the flowers blossom, and little babies to cuddle in their daddies' arms. Then came the knock on his office door and the news of the accident. He described the drive to the hospital and the hours waiting for his wife to get out of surgery, knowing already that his daughter had died. Then there was the final news, that his wife didn't make it either. He told me about the funeral—and about the years that followed. *Running.*

I started crying from deep within my guts; I had kept it in as long as I could.

"It sounds like you're running too," he diagnosed.

He didn't press much further. He knew he had opened the door, and that was enough for now. The days following our dinner brought something new into my life: I cried again. I hadn't cried since the days just after Mom's funeral. When I built my new pretend world, I had stuffed the grief down, bolted the door tightly, and made believe it was gone. His story had unlocked that bolt with a word: *Death.* Sudden death is part of this life. It happens. We can't run from it. All our pretend worlds inevitably collapse. We have to face it, ask our questions, and do everything we can to find the answers—before it comes again.

The grief was all there, just where I had left it three years before...waiting for me.

The Real World

I stopped running, looked into the mirror, and hated what I saw. I had never come to terms with myself. On the outside, I was an innocent victim. I had lost my home, my family, and my mom. Death had taken it all away. I was put in a boarding school at sixteen and forced to fend for myself...deal with grief...fight the loneliness...find a way to survive. It wasn't fair. I didn't deserve it. I was an innocent victim!

On the inside, I was guilty. I knew it, and I started to own up to it. When Mom got sick and died, I grieved for her. But I grieved for myself, too. I didn't want to admit it, but it was there. I blamed Mom for my suffering, for my loss. And now, I blamed her for the years of running. Her death had forced me out of the perfect childhood where I was loved, protected, and free. It ended forever a life without worry, without the weight of death wrapped around my heart.

I felt the anger and I hated it. I would rather have remained the innocent victim: *All this happened to poor me.* But the truth stared me down: *I'm a self-centered man. It's always me first.* Did Mom choose to die? Did she do it to harm me? Was that her plan? Of course not. So why was I angry? Is it possible that I blamed her because I couldn't control my own feelings? Like the day I was sent to camp and felt so homesick. I cried then, weeping for myself. I was scared to feel the loss of my home—loving my family, missing them, and angry that they had done this to me. I was like the little child who stomps his foot when he doesn't get his way.

When I recognized my tears for what they were—self-centered tears—I knew it was time to stop blaming my mom. It was time to own it for myself—to accept that I am a sinner, out for myself, afraid of any place that's not safe, and angry with anyone who would put me there. I was frightened of death—and of sudden death more. I was tired of feeling empty, burying myself in my studies, and running. And I was unable to commit to anyone else out of self-protection. *If I love, I risk. It might be my wife or four-month-old daughter in that car wreck! And what then? What will happen to me?*

Me!

Self-centered me—a sinner, not a saint. I didn't know the man in the mirror. Who was I to want to serve God as a priest? I hated Him for what He had done—taking Mom, ruining my life, and making me live without a guarantee of tomorrow. I was angered that He had made me unsure of myself and of what would happen when death came near me again. Would He hear my prayers then? Would He rescue me? Or would I die—and what happens at death? That was the primary concern of the man in the mirror: Survival. He was driven by self-concern, and it was sin! Sin said to God, "I want it my way! If not, I'll run and hide. I'll close my eyes to this unfair, unjust world, and I'll play the innocent victim."

I hated what I saw. All those years in church, the word *sinner* had meant somebody else. I pictured dirty people roaming the streets at night, pimps doing drugs, robbing the rich and cheating the poor; sex-fiends poring over pornographic magazines, swearing, lying, and drinking at the bars until daybreak. The word had never fit me! I considered myself a good man—moral, upright, and Christian. Sin was a terrible act like breaking the Ten Commandments, which I thought I had never done. *I was not a sinner.*

So I thought until then. Now I saw myself as a sinner and I loathed it. Yes, me. I felt it. I knew it. I despaired it.

Something Has to Change

The misery in my heart must have shown on my face. After class one day, a guy named Steve invited me for coffee. I didn't know him very well. On a couple of occasions, he had asked me to go to church with him, and I had declined. But I knew him to be a Christian man and a fellow classmate. A cup of coffee seemed just right.

I held my cards close to my vest that day and didn't tell him much.

"I don't want to pry," he said kindly, "but it sounds like it's been a pretty tough time for you. About four years ago, my wife left me. It was the worst experience I've ever gone through. A friend of mine stuck by me. He prayed for me. Finally, he took me to his church. Now, I had gone to church all my life, but I had never been to something like this before. Quite honestly, it changed my life."

"I don't understand," I said, listening intently.

"The people wanted to be there! They sang the hymns like they meant it. And the pastor, when he got to the pulpit, preached from the Bible as if he knew Jesus Christ personally. Well, I went back. The more I heard, the more I realized I was a sinner. I needed Christ. I needed Him to save me."

I didn't get it. I knew the word *sinner.* I didn't know the word *save.* I remained quiet.

"Look, maybe I've said enough. What about Sunday night? Are you free? There's a movie on the life of Jesus of Nazareth at my church. Why don't you come? No big commitment. It's just that I think it might help you. It might give you some answers. I have an extra ticket—it's going to be crowded."

He took out a pen, wrote down the church address, then "Sunday night at 7" and his phone number, and then handed it to me with the ticket. I was still guarded and made no firm commitment. He was incredibly polite as he shook my hand and asked me to think about it. As we parted, I fought suspicion. Did he think my questions had answers? The sin in me and the emptiness I felt weren't going away. They were part of life. No movie could solve my problems. On the other hand, he seemed so confident, so genuine. I held the ticket in my hand.

Sunday came and I couldn't decide. Back and forth, I went. By late afternoon, I had completed my school work. Finally, I decided to go late and slip in the back. If I didn't like it, I could leave just as easily with no one seeing me. I found the church and the parish hall. The room was dark and set up like a movie theater. I was surprised to see the place so crowded. A woman took my ticket and pointed to a row of chairs in the back. I could sit by myself, just as I had hoped. I was still guarded. The movie had just begun.

"Follow Me," Jesus told the man sitting behind a table full of money. And he did, just like that. This bothered the religious leaders—stuffy men dressed in long robes and proud of their high positions. The man who followed Jesus was a Jew, a tax-collector hired by the Romans. Before following Jesus, he was considered a turncoat, who betrayed his own people by taking their taxes, giving them to the Romans, and getting filthy rich because of it.

"Why do you eat and drink with the tax-gatherers and sinners?" the religious leaders asked.

Jesus answered, "It is not those who are well who need a physician, but those who are sick. I have not come to call the righteous but sinners to repentance" (Luke 5:27-32).

It started again—that deep crying, from down in my guts. My eyes were fixed on Jesus. He heard the cry of those who called out, "God, be merciful to me, the sinner!" (Luke 18:13). He was moved by two blind men who cried, "Lord, have mercy on us, Son of David!" He said, "What do you want Me to do for you?" They said, "Lord, we want our eyes to be opened" (Matt. 20:29-34). He went to them. He loved them. He healed them.

A prostitute fell to the ground in front of Jesus, weeping for her sins and wiping His feet with her hair and her tears. Jesus looked with compassion into her eyes and said, "Your sins have been forgiven" (Luke 7:48). And later He said, "The Son of Man has authority on earth to forgive sins" (Luke 5:24).

I knew the feeling of weeping for sin. But I didn't know His love for the sinner—or that He heard sinners when they cried out. Did He still? Would He hear my crying today? Would He forgive my sins? People two rows in front of me passed back some tissues. I was embarrassed, but I took them. I wasn't going to leave now.

I watched as the Roman soldiers beat Him—saw the nails piercing His hands and heard His cry. Then the nails went into His feet. They hoisted Him into the air beside two thieves who had already been crucified. Blood ran down His face from the crown of thorns forced on His head. The crowd stood by, watching—waiting for His every word. One man, looking up at the cross, began to quote from the Old Testament prophet Isaiah:

> All of us like sheep have gone astray,
> Each of us has turned to his own way;
> But the Lord has caused the iniquity of us all
> To fall on Him (Is. 53:6).

In that moment, I caught it. I had gone astray. I had sinned. But He came. He took my sin, my iniquity, upon Himself. It fell on Him. His bloodshed on Calvary was for me. How many times had I been in church, received communion, and heard the words, "This cup is My blood for the forgiveness of your sins"? It had never made sense—not until now. It was clear that my sin—even mine—could be forgiven.

The movie ended with Jesus appearing to His disciples, risen from the dead and alive forevermore! He held His followers close. They were to be His witnesses, to tell the world the good news: *All who repent of their sins in My name will receive forgiveness for their sins.* Then He smiled, looked directly into the camera, and said, "And lo, I am with you always, even to the end of the age" (Matt. 28:20).

I shot out the door. I couldn't hold it in anymore. Down the street, up the hill, into the town park I went. I was sobbing, wailing as if I had never known pain before. Nothing had ever hurt this much. It was all there. I wept for my sin, for my self-centeredness. I wept for the little boy, who had kicked and screamed, not getting his way. I wept for Mom, her sickness, her last months, her death—and finally, for the loss of our relationship.

The pain eased. I looked into the starry Ann Arbor night and said, "Lord Jesus, have mercy on me, a sinner. Forgive me. Forgive all my sins." The words from the movie came back to me: "I am with you always, even to the end of the age." He was with me? I don't know how. I don't know why. But suddenly, I knew I was a Christian. He had forgiven me...forgiven all my sins. And I felt the good news come into my soul. He was alive—risen from the dead and with me. He was with me.

I still didn't have all the answers to my questions. I didn't understand death, sudden death, or why moms and four-month-old daughters die. But that night, it didn't matter. Three years of bitter anger against the Lord and against Mom, as well as a lifetime of self-centeredness and self-worship, were forgiven at the cross of Christ, which had existed two thousand years before. I felt clean inside—like a new man. The Lord Jesus had heard the cries of a sinner once again. And He came that night, in His love, and saved me.

Constance

Marty kissed her mother on the cheek and promised she'd get out of the hospital tomorrow.

"Mother, the doctor wants you in one more day for observation."

"I don't like it. I feel better, and I'd rather be home tonight in my own bed."

"I know," Marty said. She quickly changed the subject. She was headed to a local department store. A suit she wanted was on sale. "Thirty-five percent off! Can you believe it?" Then she'd be off to dinner with some friends, home for the night, and then back to the hospital in the morning to pick up her eighty-year-old mother. The plans met with Constance's resigned approval. With one more peck on the cheek, Marty was gone.

At the end of the night, Marty got into bed and said her nightly prayers. She couldn't have been prouder of her sons—two handsome boys who were grown and vigorously pursuing their careers. One lived in Los Angeles, the other in Washington, D.C. She looked at their pictures, once more reminded that the years after her divorce from their father had been hard. She never thought she'd move back to Pittsburgh, let alone live with her mother again. But Marty, in her mid-fifties, had made a new life for herself. She was active at church, involved in community projects, and busy with her friends and caring for her mother. The boys were a bright spot—they called her, visited when they could, and dropped letters in the mail regularly.

Marty shut her bedroom door. She cracked open her first-floor window for some cool, fresh, November air. She turned off the light, and went to sleep—a deep sleep—until the dark, early morning hours.

Suddenly she bolted out of bed, awakened by the sounds, the smell. Still drugged with sleep, she opened the door of her bedroom and rushed out into a wall of smoke. Something was on fire. The smoke detectors were loud, piercing. She didn't turn around. Did she try to put out the fire? Did she think her mother was in the back bedroom? Marty must have made her way toward her. But when did she realize that Constance wasn't there? In those few seconds, she must have known she wasn't going to make it, but she tried to save herself. The smoke was too much, the fire too hot. She passed through the dining room and got to the window before she collapsed.

There was nothing left of the house. News stations from all over Pittsburgh carried the story live. The doctor ordered Constance's television and phone disconnected until the family arrived. She was up early, packed and ready to go home the moment Marty would walk through the door. She waited past lunch as her confusion grew. "Why are my TV and phone turned off? Where's Marty? Isn't anybody going to visit me today? Why won't you give me a straight answer?" The nursing staff did the best they could.

The family couldn't assemble until just past three. Erilynne and I met them in the lobby of the hospital. We took the elevator up to her floor and walked into Constance's room together. She sat in her chair by the bed, dressed to go home. She was surprised to see everyone, but Constance was a gracious woman—welcoming, witty, and genuinely wanting to hear all our stories one by one.

"Constance," her nephew began as he stooped down by her

chair. I joined him. Erilynne held her right hand. The family drew close around her. "Marty is dead."

We told the story once, then again. It was too much for her to take in. She kept looking at us, as if the wind was knocked out of her. "Marty? Marty. Marty is dead? Marty? She was just here last night." Slowly, the news took hold and the tears came. "I should have been home last night. It should have been me, not her. Not my Marty! I should have died in that fire. Marty? My Marty." I kept watching her face. This woman was near the end of her life, facing this most grievous loss—a parent burying her own child. Constance's home was gone, as well as all the possessions she had gathered over a lifetime. But a thousand times more important, her daughter was gone. Her Marty.

"It should have been me."

Our bishop came to Marty's memorial service.

"I can't sleep at night," Constance told the bishop before the service. "I keep having nightmares. I see Marty waking up, opening her bedroom door, and the fire! She walked into the fire!" She cried in his arms. The bishop consoled her, and there, with people gathered around, he prayed for her.

But during the service as the bishop was speaking to a packed church—wall to wall with people who loved Marty—he stopped. He walked toward Constance. He spoke to her as if she was the only one in the room. He seemed to know exactly what to say.

"Constance, your nightmares have to stop. Marty knew Jesus Christ as her Lord and Savior. She served Him faithfully. That night when she opened the door—when she saw the fire—I tell you, it wasn't but a second until she saw His glory! He promised in the Bible, 'I go to prepare a place for you. And if I go and prepare a place for you, I will come again, and receive you to Myself' (John 14:2-3). You keep seeing the fire, but no more! See the arms of her Savior who received her to Himself in glory!"

"That was it," Constance told me some weeks later. "I haven't had a nightmare since. I knew the bishop was right. My Marty went to be with the Lord. And that has eased my pain in how she died."

Constance lived two more years. I never heard her complain once about the loss of her home and all her valuables: The pictures of her husband and his letters. Little gifts given long ago. The scrapbook of their two children growing up. A favorite necklace Marty had bought her. The heirlooms she wanted to pass down to her grandchildren. I never heard a word from Constance about them as she started life over. She would live in a simple apartment this time, with all new furniture and all new clothes. It didn't seem to matter.

Her loss was Marty. She missed her terribly.

Death has no boundaries. Even at the end of life when we think we've run the race, fought all our battles, and deserve a restful, peaceful end, something tragic and unexpected comes. Death takes a beloved child within a second's notice. It doesn't seem fair. It doesn't seem right.

But Constance quietly triumphed. She knew what was important. Nothing could say it better than the one thing Marty left behind—the one thing that wasn't touched by the fire. Marty had left it at church on a hanger. We found it suspended from a fine leather necklace. She'd wear it when she served with me at Sunday morning service. It was her last possession, the only thing she could pass on to her children. How simple, but it was everything. It was what I'd been looking for all my life—the one answer to all my questions.

A wooden cross. Made in Jerusalem.

> Lord, make me to know my end,
> And what is the extent of my days,
> Let me know how transient I am.
> Behold, Thou hast made my days as handbreadths,
> And my lifetime as nothing in Thy sight,

Surely every man at his best is a mere breath.
Surely every man walks about as a phantom;
Surely they make an uproar for nothing;
He amasses riches, and does not know who will
gather them.
And now, Lord, for what do I wait?
My hope is in Thee (Ps. 39:4-7).

Part III

Staying Inside

Chapter Eight

Inside, Outside

I woke up Saturday morning grieving. It felt as if somebody close to me had died. But there was no face, no name. I didn't know anyone on Flight 427; I didn't know their families and friends. The dead child I had seen was not mine, but I felt like he was. I had finally crossed the line.

A pastor deals with death all the time. The phone rings, someone has died, and the family wants me to do the funeral. I go to their house as soon as I can. Sometimes I don't know the family. Perhaps they came to church once years ago and regard me as their pastor. Now their loved one has died. I consider it my duty to spend as much time as possible with them in the days ahead, at their home, and during the wake. I cannot simply perform a funeral service. I must know something about them and their deceased family member. This is the time to be with them, to pray with them, to help shoulder their loss and be a witness of Christ's love. Nothing is worse, in my opinion, than a funeral service where the pastor knows nothing of the family, or their deceased loved one, or their loss. It shows. It's mechanical, superficial, cold, and comfortless.

At such times, I am both a pastor and an outsider. As close as I try to get, I'm not a mourner since I didn't know the deceased personally. To be with the family and conduct the service as if I did is acting—and acting never comforts. But it's right to stand just outside the circle of mourners, never crossing the line, never pretending I'm on the inside suffering loss when I'm not. God's love is always sincere, always genuine. To bring His love, I must be the same. Anything more, anything less, misses the mark.

It's a different story altogether when death comes to a parishioner I've known and loved. I remember one woman in her early sixties who was ill for many years. She'd been in and out of hospitals. I knew her and her family well. When her time came to die, I was at her bedside in the hospital. It was late at night, and most of her family had gone home. One minute she was awake—chatting away like nothing was wrong. The next minute she was in a sleeplike haze. As the night went on, she slept longer and harder. When the time seemed near, she surprised us. She awoke suddenly. She looked straight at me and said clearly, "I am so sorry it's taking me a long time to die. You need to go home and get some sleep. It's late!"

She squeezed my hand and smiled. Such love! Fifteen minutes later, the heavens parted and she went home to be with the Lord Jesus Christ, whom she loved so dearly. At her funeral, I was her pastor, her friend, and a fellow griever with her family. I was inside the circle, missing her.

I had all those same symptoms that Saturday morning after the crash. Grief had come. I didn't want to follow the normal routines. Every morning, Erilynne and I have time for prayer.

[illegible]s, take our forty-minute walk, have breakfast, shower, [illegible]d off to work. But not that day. Why do routines feel so meaningless during times of grief? Miss our prayers? Two days before, I couldn't imagine it. Now, I didn't care. The phone rang; we let it ring. The dogs needed to be walked. *Why can't dogs walk themselves?* It's time for breakfast. *Who wants to eat?* Nothing felt right. What had meant so much before meant so little now.

It's even hard to have a normal conversation at times like this. An outsider can talk about the weather or a lengthy battle with the flu. On they go, talking away. But I find it difficult to stay focused. The grief hurts too much. And the grief is all I want to talk about—all I want to think about. It is, for that time, all there is. The world feels upside down. Everything is seen through the lens of loss, and it all looks so different—family, friends, job, money, possessions, plans in the present, dreams of the future, what I did in the past, what I should have done and didn't do. Call it an obsession, but grief is like that. It rules the heart, demands the mind, and captures the soul.

I had crossed the line.

As much as I wanted to go back to the crash site, I couldn't go until noon. I had to spend the morning preparing the Sunday sermon. I called Ken. We agreed to meet at the Plaza at twelve o'clock sharp. John couldn't go. He and his wife were leaving on a six-day preaching mission. I went to my office, sat at my desk, opened the Bible, turned on the computer, sipped my coffee, and stared blankly out the window. The sermon I had been preparing during the week wouldn't work. Not now, after the crash. I had to start again. But once more, it was hard to keep my mind from wandering.

I began to relive the events that had occurred since Thursday night, step by step. I remembered the interviewer's question on live television, "Tell me, why did this happen?" I would have no choice in the sermon but to ask the same question and to answer it simply and directly. Our beloved

Prince of Peace Church would be grieving. There were pilots and flight attendants in the congregation as well as USAir management personnel, mechanics, ticketing agents, air traffic controllers, United Airlines employees, and people who worked at the airport. Some knew people who had been on Flight 427. Others saw the plane go down. All of us were reeling from the shock.

The service, I thought, would be like a funeral. I was starting the sermon outline when the phone rang.

"Guess what?" The voice on the other line was Joyce, our church administrator and good friend.

"What?"

"Two calls came in this morning. ABC local news and CBS national news are coming to church to film our Sunday services. They want to hear what you have to say."

"They want more than that," I came back sharply. "They want to see our pain! What could be better than to catch the tears of our USAir families at church while they're praying, then afterwards, to interview them?" My emotions were right at the surface. I apologized and thanked Joyce for her call. I knew I was reacting to my own grief. It's typical for insiders to find reasons for getting upset. It gives us a way to vent the pent up emotions steaming from within. For example, I met some who were angry with USAir, others were mad at Boeing, who had made the plane. Not me. I had found my own personal outlet: the media.

The Media Hawks

Mid-morning, the phone rang again. A newspaper reporter wanted an interview.

At first, he wanted background information: Where was I when the plane went down? Why had I gone to the site? Who went with me? What was my job? Was I going back? Naive me, I thought the man was interested in a chaplain's work. I told him plainly, "The Lord promises in Psalm 23 to be with us in 'the valley of the shadow of death.' He doesn't stand outside, watching from a distance. He's there in our suffering. The rescue workers needed to know that. That's why I went," I said passionately.

He wasn't interested. "That's fine," the reporter said coldly. "Now what did you see up there?"

My stomach turned. He wanted raw descriptions of the bodies. That was his story, and I wouldn't give it to him. He tried a different angle: Have you ever seen violent death before? What was it like for you? How were the rescue workers? Were they breaking down? Is there a story of a worker you helped who just couldn't hack it?

I tried to remain polite. I told him that information was private.

"What newspaper are you with again?" I asked bluntly. It felt like I was being interviewed by one of the afternoon talk shows or a grocery store rag. I could see the headline: "Chaplain Tells All! Violent Death Described in Detail! Rescue Workers Leave Site in Tears!" They wanted titillating stories to thrill the public into buying their newspapers. How could anybody make money on this human tragedy? The whole thing made me nauseous. It got worse when I learned the reporter was a religion writer for a respectable, conservative newspaper. He pressed again with more questions, but I couldn't continue.

Once again I went back to the sermon. Once again the phone rang.

It was a woman in New York City. She worked for one of the major TV networks. Her job was to line up guests for their morning news program. She wanted me on their Monday show.

"We're looking for a new angle on this crash story in Pittsburgh," she said with fascination in her voice. "Our producer wants someone who's been at the crash site, preferably somebody from the human services end like yourself—a clergyman. You've been at the site, seen the demolished plane, seen all those poor people who died—how could you stand it?—and worked side by side with the rescue workers. You've been there. Then at the same time, right next to you, we want to interview a family member. What a contrast! You've seen their dead. They haven't. They won't. All they can do is mourn their loss. And there you are, a clergyman, comforting the family member on national television—live!"

"That is revolting!" I said, unable to control my reaction.

"You won't do it?"

"I will not. I'll have no part in it."

"Well, what about you alone?"

"I don't think so."

"Why, uh," she stumbled, still persistent, "why don't you just tell your own story?"

"I have only one story. I'm a chaplain serving at the crash site. That's it. My job is to be with the rescue workers—to comfort them, to tell them the Lord is with them as they care for the dead, and to pray for them. I cannot—I will not give horrid descriptions of the dead, nor will I tell stories about individual rescue workers. That's private. You're looking for fireworks, and I can't help you."

"Sure you can!" she said happily. "Let me call you back. We'll set up the interview time and the location for Monday morning. Our limousine will pick you up at your house if that's okay." It wasn't long before she called back. The producer didn't go for it; I didn't have the spice they needed.

I turned on the television. The local news was still covering the crash, but it seemed different to me. Early on, the media had wanted the facts: What happened? What was being done? Who was in charge? Did anyone survive? Were there eyewitnesses? Even on Friday, the media seemed to be at the leading edge of the latest news. But sometime that afternoon, it changed. The flow of new information slowed down. The NTSB, FAA, and the county coroner had nothing new to say. The cleanup effort was in place and it would take a week or so to complete. Any new developments would be reported at the daily press conference, not hourly. That was it.

That was hard for the media to take. They wanted to keep up the frenzied pace. They needed stories from other sources that were just as dramatic—and they found them: The man who had a ticket for Flight 427 but, who, at the last minute, decided not to go. A mother whose son died in the crash. Stories from business workers remembering the dead. An eyewitness who was sure the plane hit a flock of birds. Another who said he saw an explosion before the plane dove into its fatal spin. *A bomb!* This gave rise to a rumor filled with mystery and intrigue. Evidence suggested that the FBI had a passenger on board who was part of their Witness Protection Program. Was he the problem? Did someone want him dead? *Sabotage!*

I turned the TV off and sat at my desk. It would take a few days for the NTSB to formally rule out both theories—no birds, no bombs. But right now speculation ran wild. No one knew why this plane had crashed. That gave rise to more people with more theories. And, of course, that meant more news coverage.

"Are you the Reverend Barnum?" The phone had rung again. It was another reporter looking for another provocative news angle for the outside world. I didn't give in. He pressed. He wanted to meet me at the Plaza.

"Let me ask you a question," I urged. "Where are you from?"

"New York City."

"So you flew into Pittsburgh?"

"Yep. Yesterday morning around eight-thirty."

"And what was it like? Were you scared? Did you think about the crash? Did you think it might happen to your plane? Was it different this time? Could you feel what those people who died felt?"

"Yeah, I was scared. Landing felt good. I was pretty relieved. I couldn't help thinking about it."

"And did you pray? Did you think about dying? Do you

know what happens to you at death?"

"I'm not a religious man," he confessed frankly.

"Maybe it's time to start," I said urgently. I told him life was short. We don't have much time on this earth. We have to get right with the Lord now, before it's too late. He listened, half tolerating my appeal. I think it proved one fact clearly—he knew I wasn't the right man for his story.

It also proved I was on edge. The sermon wasn't coming together. The morning was almost over. And I couldn't stop reacting to the media's lust for the sensational story. The suggested interview on national television had disturbed me. I imagined the show's host asking an insensitive question, making the family member cry. All the raw, uncontrolled emotions of loss would rush to his face. Suddenly, he wouldn't be able to talk. He would hold his head in his hands, sobbing. Then he would try to regain his poise for the camera, but it wouldn't work. The camera then would zoom in on the clergyman. He's been to the site. He's seen their dead. He wouldn't be able to control his emotions either. He would move toward the family member, and they'd fall into each other's arms weeping, as if they were the players in a Hollywood script. All our grieving would be on public display for the world to see.

Outsiders watching insiders.

It felt vulgar to me. Why can't the media stand outside our circle? Why can't they leave us alone to mourn privately. But they don't. They crash through the line—microphones, lights, cameras, and reporters asking inane questions. Nothing is private, nothing sacred. Insiders don't matter. They steal our tears for outsiders to see. They do it for money, because it's their job. Looters! Isn't there anyone who protects the insiders and decently upholds our honor? What about the police who guarded the crash site on the first night? Looters came to rob the dead. The police caught them. How horrid, everyone thought. But what about these looters who steal a story from our broken hearts, who make a buck at our expense. Isn't there anyone to help?

"That is revolting!" I told the newswoman.

Then a scene flashed through my mind. Only two weeks before, I'd committed the same crime. I was a guilty outsider feasting my eyes on an insider's tragedy. I had stared without thinking.

Erilynne and I were on the Pennsylvania Turnpike; our car had slowed to a crawl in heavy traffic. For twenty minutes, we inched our way forward until we saw, in the other lane, fire trucks, police cars, and two ambulances. The closer we got, the worse the scene became. There were two cars. One was on its back, belly up, still steaming from the fire. The other was smashed in the front, with its windshield gone and the driver's side door standing open. Medical teams were working on the victims. What appeared to be family members were huddled together around the car. The closer we got, the more we saw and the more intently we looked. We saw faces anguishing from pain, crying… bodies lying helpless, fighting for life, bleeding.

As we passed by, a victim was moved to a stretcher and carried to the ambulance. The sirens went on. The back door closed, and the ambulance took off. In front of me, the cars picked up the pace. Within seconds, we were driving at a normal speed. There was no construction in our lane, no accident to slow us down. The traffic had been backed up for half an hour due to outsiders who wanted to see the accident, catch a glimpse of the blood and wreckage, and then drive on. It's called rubbernecking.

Erilynne and I prayed for the victims and their families. We talked about it for a while. Later that day, we told the story a couple of times, saying, "We saw a terrible car accident on the turnpike." But we soon forgot the event. We stopped praying for the people. Our lives went on as if nothing had happened. Not once did I think it was wrong to stop and stare. Not once did I feel for the family members who, looking across the highway, must have seen us rolling down our windows to gawk at their personal tragedy.

Why do we do it? Is there a place for a natural, healthy curiosity and concern? Is there a line we cross where we violate the privacy of people who are suffering? The movie industry, in recent years, has discovered people's hunger to see violence, bloodshed, and scenes in which all is bared. The movies of yesteryear held the line. They had a certain decency, they knew what to show and what not to show. That sense of decency is gone today. Money-makers have found a secret about us: We like to stop and stare. We're rubberneckers who'll pay to see it all.

That's what I did on the turnpike. I was guilty. I didn't see the incident from the insider's perspective.

It was hard to recognize my own guilt and stay angry at the media at the same time. I still protested their inability to hold the line of decency. Outsiders didn't need to read or see certain stories at the cost of an insider's pain. But for me, I had to let it go. I had work to do at the crash site. I had a sermon to prepare and a church to comfort in a terrible time of tragedy. And I had my own grief, which I didn't need to take out on the media as if they were my personal punching bag. It was better to deal with the grief straight on, with the Lord, in prayer and over time.

I didn't finish the sermon until after dinner that night. The morning had passed, and it was time to meet Ken at the Plaza. I turned off the computer, stood up, and knew one thing for certain: I needed to have more compassion for outsiders. They didn't understand. They weren't suffering like we were. This crash wouldn't touch their lives in the same way it would ours. Somehow, I needed to accept that and keep going.

Ken: An Insider

The emergency command headquarters had moved its location. The Green Garden Plaza was simply too congested with shoppers and rescue personnel. Gradually, the rescue operation moved across the street from the Plaza, where there was less confusion.

I met Ken in the Plaza parking lot. It took us a few minutes to realize the change. We crossed the road, found the trailer for the decontamination suits, and signed up. It was nice to see familiar faces again. Relationships build fast in a crisis. People are usually more open, more trusting, and more vulnerable than at other times. We talked with some of our new friends before putting on our suits and boarding the bus.

For a split second while Ken was dressing, I saw something in his face. Was it insecurity? Did he not want to go back? Was I pushing him too hard? Was yesterday too much? Why hadn't I talked to him about it? I kept forgetting that Ken had just graduated from seminary in June. He was newly ordained and in his first position as an assistant pastor. A man of thirty, Ken was married to Sallie, a gifted lay pastor in her own right. They had two boys, Mark, five and a half, and Will, three and a half. How could I expect him to serve at the site? Was my confidence in him too strong? Was I presuming too much based on our friendship? We were kindred souls, closer than brothers, and a good team. What was going on inside him?

"How are you doing?" I asked him once we were on the bus.

"Last night was hard."

"What did you do?"

"Sallie and I talked for a while. I took some time by myself—wrote in my journal, prayed. It was easier being at the site working than being at home, thinking about what we saw."

"Did you sleep at all?"

"Some. On and off."

He looked out the window. We were now on our way back up the hill. His eyes filled with tears.

"Seeing the dead child," he said, his voice breaking, "I couldn't shake it. I finally went into Mark and Will's room late. They were both sleeping. I knelt at Mark's bed first. I just had to put my arms around him. I wanted to hug him so tight. That's when I lost it. I started crying. Everything in me wanted to protect him from harm—to keep him safe from what happened to that dead child. I couldn't let him go. I knew I had no control over his life. I guess I had known that before, but it was different. I felt so helpless as a man—as a father. I don't want my boys to die. Not ever. And not like that. I went over to Will's bed and the same thing happened. I couldn't stop crying."

I put my arm around him. Death had come, and he was standing in its wake. It felt as if his own flesh and blood had been ripped from his side. He spoke the truth when he described the feeling as being helpless, with no power to save his children, his wife, or those he loves—not even the power to save himself. He had no power to stop the forces in this world from coming near, and no ability to pretend death isn't bigger, stronger, and larger than life. There was nothing to do but cry.

The bus stopped in front of the Salvation Army canteen, just as it had the day before.

"Are you sure you want to go?" I asked, trying to give him an out.

"Yes, I'm sure. When the Lord called us to serve at the crash site, He didn't say we'd be free from pain. I knew that. He said He'd be with us. And He has been, and He will be. I want to be there."

"Ken, I can't imagine what this would be like without Him."

"Nor can I."

He stood up, resolved to go back to the site. We got off the bus and walked down the logging road toward the tent. It was different from yesterday. Actually, I was different. I didn't have the same butterflies in crossing the line where the yellow tape had once hung. I didn't have the shyness or the lack of confidence in my work at the site. I didn't feel like a stranger peering from a distance at this dark, miserable world. The change surprised me. It seemed right to be back. It felt like I belonged there with the people who were experiencing the crash site firsthand, there with all the other insiders.

There was a noticeable difference that day. The smell of decaying bodies was now very present. I had never smelled anything like it before. It was extremely pungent, and it was everywhere. We couldn't get away from it. I don't know why we didn't put Noxzema creme in our noses that day, but we didn't. We tried to use the mask, but it didn't block the smell. It was there—part of the site and part of us.

Death's stench made the sight of death less tolerable. The two combined made the second day just as difficult as the first. Ken and I spent the first hour talking with rescue workers we'd met the day before. Conversation came easily—we just picked up where we had left off. But it was clear that the shock wasn't going away. Most of them told us they had thought it would be easier the second day—not that they'd become callous, but they had hoped for some sense of relief. It wasn't there. That was frustrating for them and for us.

At one point, I looked up at the impact site. I saw rescue workers I hadn't met before. I saw a friend, the Reverend Carl Neely (my heart leapt to see another chaplain at the site!)—a godly man who holds true to the Bible, an Episcopal priest whose first career was in the military. I also saw Wayne Tatalovich talking on his portable phone. A small group of men and women surrounded him. I wanted to be up there. I was ready, but it wasn't time. We were in the middle of a conversation with a rescue worker. Also, I wasn't sure if Ken wanted to go to the impact site yet.

"Will you guys pray for me?" the rescue worker asked us. He needed the Lord's strength to do his work and to cope with the sights and smells. There, in the middle of the logging road with workers all around us, we took his hands, bowed our heads, and prayed for him in the name of the Lord Jesus. It was, he said, exactly what he needed to survive the day. He smiled warmly, shook our hands, and left us. Once again, I looked at the impact site. Maybe now was the time to go. I turned to Ken.

Surprise.

He was gone. He had seen a rescue worker some fifteen feet away. The man had stopped digging and just stood there, frozen like a statue and staring into space. He looked friendless and alone. I watched Ken make his approach, and in just a few minutes, turn this man's stony face into a soft one full of expression. I marveled that Ken had stepped out on his own. I felt a certain victory for him.

With that, I went to the impact site by myself.

"You were my yellow tape" he told me later.

"Your what?"

"You said you felt safe outside the yellow tape. You didn't want to cross it, but you knew the Lord was calling you. You had to step out and trust Him. Well, I found that same safety in staying at your side. It was easier to follow your lead. For me, the destruction of the plane, the bodies, the smells, some of the despairing looks on the faces of the rescue workers were overwhelming sometimes. I didn't think I could be there by myself. You seemed confident, and I didn't have that. So I didn't leave you."

"What happened to make you leave?"

"As I watched you talk with the workers, I saw how much it meant to them. Slowly, the idea started building inside me: *I need to do that, too. I need to go on my own.* The more I stayed next to you, the more I felt the Lord prodding me to go. And that was scary. I didn't want to leave you or the comfort I felt. Then we prayed for that rescue worker. We asked Almighty God to give him strength to do his work. I thought, *How can I pray that prayer for him and not for me?* I needed that same empowerment. That's when I saw the worker standing by himself, looking so lost. My heart was moved for him. And I realized my concern for that man was bigger than my fears of being alone or of being overwhelmed by the crash site. So I made the move."

"And it worked?"

"It did. The man could have brushed me off and sent me right back, but he didn't. He was very receptive to me. He wanted to talk. Halfway through the conversation, I realized I had the confidence I had been missing before. I had the strength to do this work, and that amazed me. I had known the Lord was with *us,* and now I knew He was with *me*. I wasn't alone. By God's grace, I'd left your side and crossed my own yellow tape!"

"And you still felt safe, right? That was my experience."

"Yes, that never went away. I can't say it was always smooth. Some of the workers wanted to talk to a chaplain. Others didn't. I remember one man—the struggle was written all over his face. I went to him and said, 'Are you okay?' At first, he wouldn't look at me. When he did, he wouldn't talk to me. It was awkward and frustrating. Normally, I'd just leave. But I wanted him to know I was praying for him. It looked like he was trying to handle it all on his own. Somehow, I wanted to break through.

"It happened for one brief second," Ken continued. "Two days later, I saw him again. He was with a coroner's team, standing over some remains. Our eyes met. I saw his eyes water and a look of despair come over his face. He knew who I was and that I was praying for him. He opened his mouth to say something, as if all his defenses had dropped. But he stopped, and turned his head away, and I could tell the defenses were back up. I said quietly, 'I'm here. And I'm still praying.' He didn't react. I saw him many times that week. His face was so hard. I wanted him to know the Lord. I kept praying that his heart might soften so that he wouldn't have to carry all this grief and tragedy alone. I ached for the man."

"Praise God you were there for him."

"I wouldn't have been if I had stayed glued to your side."

"Yeah, but you didn't. You crossed the yellow tape and felt the power of the Lord to grace your way. You knew His strength by experience, working deep inside you. Now it was your job to tell others about it, so that they might trust the Lord in their grief and find the same strength—the same comfort you found."

"And that's what I did at the crash site," Ken said. "But taking that step away from you had its negatives, too—not in serving the rescue workers, but afterwards, when we left the site."

"When you were by yourself?"

"Yeah. Saturday night was hard for me. When I left your side, I knew I had a purpose up there. But when I came home, it was gone. I sat down with Sallie and tried to tell her what the day was like—how I felt, how I was dealing with the grief. There was so much inside me. But I couldn't share it. It didn't have words yet. I didn't know what to do but cry. We prayed together and held each other. I realized then that I needed Jesus Christ to be with me at home just as much as I needed Him at the site.

"That night," he went on, "I still couldn't escape the smell. It was on my clothes, on my skin, in my hair, and in my nose. I had to take a shower and wash it away. But it didn't work. The smell was still there. All that night and through the church services the next morning, I could smell that smell of death. My eyes would water and I'd just weep. There was nothing I could do. I couldn't get away from it."

"I couldn't either."

"Crossing the yellow tape isn't easy," he said.

"No, it makes us insiders."

And that's what we both were. We had crossed the line. We were grieving for the 132 victims and their families, and for the rescue workers, who smelled just like we did. We grieved like those who had lost their own children, wives, families, and friends. We were facing death, and there was nothing we could do to wash it away or make it disappear. It was ours; we had to live with it and let it run its course.

"I had thought I could keep my distance," Ken said flatly.

"Not a chance. Once we crossed from the outside to the inside, there was no way back."

Chapter Nine

The Impact Site

There were two ways to get to the impact site. One was straight up a hill on a small path. The other came in from the left of the site and took a more gradual slope. The Delta team had built this second way of access. It connected the area to the logging road near the Beta side. It was wider and allowed bigger equipment to get in and haul the plane parts away. I opted for the small path, climbing the hillside carefully. Rescue workers were on either side, sifting through the dirt. To my right, a coroner called for assistance. He had found another body. I got to the top and stood there, frozen in my tracks.

The area was heavily wooded, but the impact site was open. Trees surrounded it in a horseshoe shape that was about fifty feet in diameter. Directly above me was open sky. The trees on either side were scorched by fire and riddled with debris. In front of me was the cockpit, but I couldn't tell it at the time. I expected to see the nose of the plane, control panels, a throttle—something familiar. All I saw was wreckage meshed with complicated wiring. To my left, I saw one of the engines.

From my right, I heard a woman say, "Look at this!" Her eyes gleamed. She was on her hands and knees, combing the dirt with a small tool. In her left hand, she held a ring. She sat back and smiled. "This makes it all worthwhile. Some family member is going to treasure this ring forever."

"Can I see it?" another woman asked from a few feet away. "It's beautiful." The ring was placed into a clear plastic bag and labeled. The two women went back to their work. One talked about a wallet she had found an hour before. The other said she had prayed that God would help her find things like that. She thought it would bring some happiness to the families. And it made her work so important, so hopeful.

Behind them, I saw rescue workers searching the woods, mostly alone. Some were standing, raking the ground. Some were on their hands and knees. It was the same everywhere I looked. Closer to the impact site, I saw coroner teams grouped together over different spots. The area still had a lot of bodies on the ground—more than I had seen on the logging road. It took time for me to take it all in.

This was it—the place where the plane had come in at three hundred miles per hour. This is where the 132 people had died. Right above this spot, the pilots had fought with everything in them to recover the plane. They screamed to the tower for help. Then came the silence. It happened here—the ball of fire, the explosion, and the black smoke. In the sky just above me, they had lived their twenty-three seconds of terror. On the ground below me, they had died their sudden deaths. One minute in time, the next in eternity. It had all happened here.

It took my breath away. I dropped my head in silence. I couldn't help thinking about the people. How many were ready to die? How many knew the Lord and entered into glory with Him? When I opened my eyes, I saw a piece of paper under my foot. It was from a national weekly magazine. The headline was in big bold letters. I stooped down, picked it up, and read the caption: "A Destination Close to Heaven." That said it all. For some people, this flight went nonstop from Chicago to heaven's shore.

But there was an agony in my soul—wrestling deep inside. It was the question no one wanted to ask or think about. Where were the people on board who weren't ready, who didn't know the Lord, who were blind to His love, who had chosen not to be on the path toward glory? What about them? *Lord, in Your mercy*, I prayed, *what about them?*

I couldn't control myself. I wept for the lost.

The Silver Lining

Standing there, I overheard two men examining the obliterated cockpit. I watched them study the rubble like pathologists over the dead. Every few minutes, a rescue worker would interrupt them, holding a small piece of metal. The men would look at it, sometimes briefly. Then they would identify what it was, what it did, and where on the plane it had been located. I was amazed at their precision. To me, it looked like twisted and broken junk. But to them, each piece was part of the whole aircraft. No matter how small the fragment, they knew exactly where it fit—left side, right side, engine pieces, cockpit electronics, metal from the seat construction, and parts from the fuselage. These were professional men, courteous in manner and directed, as if they had a mission to accomplish.

"I'm glad they sent chaplains up here" Sam said in a southern drawl, as he looked at me. Both Arthur and Sam were thin, just over six feet tall. They appeared to be in their mid-forties.

"Thanks," I responded. I walked toward them, pointing to

the exposed cockpit. "I'm impressed. You guys know this plane inside and out. You must work for Boeing. Are you airplane designers?"

"No. Actually we're pilots," Arthur said as he introduced himself. He was from Charlotte, and Sam was from Atlanta. "I'll bet you expected to see something intact. Could you tell this was the cockpit?"

"I had no idea," I admitted. They invited me to come closer and explained the cockpit in simple terms. Somehow the position of the wiring told them where the pilot and copilot had sat, how the plane came in, and what happened on impact. As hard as I looked, I couldn't see anything recognizable, nor could I imagine how the maze of metal and wires once formed a real cockpit. It was beyond me.

"So you're here to find out what caused the crash?" I asked.

"Yes. We're part of the investigation team," Arthur replied. "The National Transportation Safety Board (NTSB) is responsible for finding out what happened to this Boeing 737-300. They call on experts from different fields to be part of the investigation. The first job is to recreate the plane piece by piece. It will leave the site and go to a hangar at the Pittsburgh airport. From there, it'll be reassembled. That's when the work really begins. The team will analyze the aircraft to find out what made it fall from the sky."

"You know, planes don't just fall from the sky," Sam interjected. "There's no safer mode of transportation than flying. There's more risk in driving to the airport than in flying to your destination. Now, I was at the crash site of a DC-9 this past July in Charlotte. It had hit bad weather and what we call a wind shear. But not this plane—there were no adverse weather conditions in Pittsburgh. No sign of difficulties. But we know something fatal happened to the aircraft, and we're going to find out what it was—no matter what."

"I can't believe you'll be able to recreate the plane," I remarked. "All these thousands of pieces?"

"It is possible," Arthur stated. "We're hoping the actual clue is here at the crash site."

"You can be assured," Sam replied with determination, "that we will find that clue, come hell or high water. The NTSB has determined the cause of nearly all commercial airline crashes over its long history—all but a few. One of them happened a few years ago in Colorado Springs. To this day, that case is not resolved. The plane went down just like this one. It dove straight into the ground for no apparent reason. Let me tell you, this case will be resolved. I'll stay up night and day through December if I have to; we're going to find the cause. Mark my words, what happened here will never happen again."

Sam was passionate, matter-of-fact, and convincing.

"We listened to the tape a dozen times last night," Arthur said.

"What tape?" I asked.

"The cockpit voice recorder. The last thirty minutes of pilot conversation is recorded. It also picks up the sounds in the cockpit. We thought maybe if we listened to it, including the final twenty-three seconds, we might hear something abnormal—anything. As a pilot, I put myself in their position. What happened up there? I want to know. Those two guys didn't have a chance. They did everything possible to bring that plane out of its downward spiral, and nothing worked. Why? Pilots are trained for emergencies like this one. We know what to do by instinct. There's no time to think. And these guys did their best. They gave their all. I'm like Sam. We've got to figure out what happened on that aircraft. It's the only way to honor the two pilots who died and to prevent it from ever happening again."

"Was there anything on the tape?" I asked, horrified at the thought of listening to it.

"Nothing conclusive," Sam reported. "But we've only just begun. The flight data recorder is at the NTSB lab in Washington right now. That will tell us the plane's heading, altitude, pitch, airspeed, all its engine functions, and so on. Right now, it's all speculation. I'm personally interested in the rudder. Some thought its malfunction was the cause of the Colorado Springs crash. The flight data recorder on the 737 doesn't monitor the rudder. So we need to find it, put it back together and examine it."

"And we'll do it," Arthur said, with confidence. "We are going to find the cause!"

A rescue worker came up beside us, holding a piece of the plane. Sam recognized it immediately and sent the worker on his way. We talked for a little while longer. I couldn't help feeling a sense of hope rising inside me. I believed them! They were going to find out why this plane had crashed! They had both the passion and the expertise to fulfill their mission. Nothing could bring the plane back, or the people. But there was one cause left—one thing to fight for. What made the plane crash? Could it be prevented in the future? Arthur's and Sam's single-minded devotion to their cause was infectious. They had found something to fight for, something in which to hope. That was a silver lining in this dark thundercloud.

For a short while, their enthusiasm overshadowed the grief I'd felt when I first came to the impact site. I saw their drive and got excited about finding the source of the problem. But it didn't take long for that to pass. My grief was for the lost who had died, not for how they had died. Finding the cause wouldn't change that. It couldn't bring them back. Yes, it was crucial work to do, but not one that eases the pain.

Grief plays tricks on us. It's always hard to face it straight on. We need distractions—someone to blame, something to fight for tooth and nail. It gives us a reason to get up in the morning. We push the grief aside and turn our whole attention to our cause. Meanwhile, we don't let grief have its way with us. It needs its time, its place. There's a season to grieve and recover from tragic loss—to find ourselves and to establish our lives again. It's far better to push the distractions away and deal with the grief. But too often we run from it, give ourselves to a cause, and rationalize, "It's the only way I can survive."

For example, when a mother loses a child to an accident involving a drunken driver, it's common for her to take a stand against drinking and driving. It's a noble cause that saves lives. But it can't bring her child back. Nothing can—no devotion to the cause, no money from a lawsuit, no satisfaction from saving one life. It may be rewarding. It may, for a time, soothe the emptiness inside. It may give a sense of hope when the heart feels hopeless. But it will not take away the personal loss or the human suffering of grief.

We must go through the grief at some time and in some way.

Before I left Arthur and Sam, I spotted something bright orange in the ruins of the cockpit.

"What's that?" I asked, bending down to look more carefully. Finally, I reached for it.

"What are you looking at?" Arthur asked, ready to identify the plane part.

I picked it up in sheer amazement. In my hand, I held a soda can—a carbonated orange drink. The can was intact, untouched and unopened, as if brand new and ready to be served by a flight attendant to a waiting passenger. I held it up for their inspection. Sam took it and spun it around in his hand. There wasn't one scratch.

"This was aboard the aircraft," he said with certainty. "It's the kind we serve."

"How could this have survived the crash?" I wondered aloud. Both men shook their heads in silence and dismay. Sam gave it back and I thought to myself, *What a picture. Nothing survived; the plane is destroyed. The people all died. Only a soda can is left in its normal state.* I tossed it back. There was no comfort to be found in it, no heartening news that something had survived. In fact, it made the grief that much harder. If this could have survived, why not just one person from Flight 427?

Nothing can bring back the dead—not a soda can, not a

hope to find the cause, not a silver lining in a thundercloud that left destruction and death in its wake.

Tell It Like It Is

"I don't want to hear it," said a man's voice behind me. It surprised me. For an hour or more, I had walked through the impact site, staying near the coroner teams and talking to the workers. But I had stopped at the crest of the hill and found myself standing alone, staring down at the logging road, daydreaming. In my mind, I was in church at the pulpit, ready to deliver the Sunday morning sermon. *But what was I going to say? How could I witness such misery and at the same time bring comfort? What if I can't do it? What if I break down in the middle of the sermon? What if—*

"I'm sorry?" I said, startled by the voice.

"I said, 'I don't want to hear it.' " He came right up to me, and stood only a few inches away from my face. We were the same height. He looked a little older than me, maybe in his early fifties. The top of his head was balding, and the rest of his chestnut brown hair was gathered in a ponytail. The sharpness in his voice was just as present on his face. The lines of his face were etched into a deep scowl; his dark eyes were piercing, almost angry. He wore an annoyed look—stern, frank, and raw. His presence was forceful, leaving me to think I had offended him in some way.

"I don't understand," I answered, not willing to back away.

"You have all the answers, don't you? I've been watching you talk to everyone, and I know what you're up to. You think you can waltz in here, say a few nice words, and make us all feel better. Well, let me tell you, Bud, it don't work like that. There's nothing you can say to change what happened here."

"I think you're putting words in my mouth," I responded slowly.

"Maybe," he retorted gruffly. "But I don't think you pastors see right."

"What do you mean?"

"All you see are struggling rescue workers who need a loving pastor by their side. You don't see what I see. Look there!" he said, pointing a dozen feet below us. A body was being brought up from the side of the hill. There was another one below it, then a third. The power, the force of the impact, had buried the bodies just below the surface. Even from our position, we could see the shapes of the bodies. By the time they got to the third victim, the coroner team stepped back for a break, as if it was too much to handle all at once. There was too much death, too much personal pain. It was hot. The bodies smelled. The end of the day was still a few hours away. One of the men swore. Another wiped his brow in frustration.

"That's what I see, preacher," he said with resignation.

We stood there together, our eyes still fixed on the scene below. The one woman on the team, a coroner, stooped down. The anguish in her face, the red around her eyes, and the tremble in her hands revealed a pain that didn't stop her. In a gentle voice, she pleaded with her fellow workers, "Let's bring him out. Then we'll take a break. What do you say?" I could see how her words brought comfort to the three men. And I knew why: *She knew the pain. They shared it together.* There's comfort in that—a melancholy comfort.

As they resumed their work, I got it. This man next to me had a message. He wanted to put my face into the face of death—and make me feel it, make me stay there until I had nothing nice to say, no cheap words of cheer to bring. What a contrast to Arthur and Sam! They were optimists, who avoided all this death by doing something positive. They were upbeat—men on a mission. But this man was behaving like a master who was house-training a new puppy. The puppy's nose is forced into the dung, he gets a slap on the rear, and the master shouts, "No! Bad dog!" He wanted to see my nose in the dung of death; he wanted me to experience it and not forget it. I was sure that was his message.

"You want some advice, Bud?" he said, not knowing my name. "Don't make it easy. Don't tell us everything's going to be all right. This place is a hellhole. It stinks. We're going to carry what we see here for the rest of our lives. So tell it like it is. Be real. Feel the pain we feel before you open your mouth and give us all that smooth talk. This hellhole is reality. Talk to us like you know that!"

He turned to walk away, his point made.

"You must think all preachers are the same." I pushed him, hoping he wouldn't leave.

"Guess I do," he said, turning back, "because I've never been to one funeral where the preacher knew anything about pain. It's always nice words, nice music—a real nice service. But it's not real at all. When was the last time you asked the hard question? Where is God in all this, huh? Answer me! You all beat around the bush. You tell us God loves us, He cares, He's always there. Well? What about now? Are you going to deal with the hard issues? We have to live with them. Does He? Does He really care? Does He have an answer to *that?*" he asked as he pointed down to the coroner team. "Or doesn't He? What does He know about our suffering, our pain, what we've gone through? Give us the answers, Preacher!"

He said it with such conviction, such need. But he wasn't going to let me talk. He backed up a few paces, began to turn, and then stopped. He had one more word for me before he left.

"Do me one favor," he said thoughtfully, "when you get up in your pulpit tomorrow morning."

"What's that?" I asked, sorry he was leaving in the middle of our conversation.

"Don't let God off the hook."

I never forgot those words. I nodded in appreciation and gave a half smile. Inside, I resolved to do just that: *Tell it like it is. Be Real. Don't let God off the hook.* I wouldn't in tomorrow's sermon—and never again. And that was that. I never saw him again. I've often wondered if he knew how much I needed that talk. Did he think, *He'll never listen to what I say?* Well, I did listen. I saw the death, I felt it deep in my soul, and I'll never forget it. Never. There would be no more "nice words" from me, but instead, real words—no more talk of a "nice God," but of a real God who has answers at times like these…real answers to real suffering, and real love for our deepest pain.

Firsthand

I didn't move. I watched Bud (I didn't know his name either) walk down the path toward the coroner team. He stood with them as the last body was carefully removed and taken away. Then he joined them as they headed slowly down the logging road, through the tent, and up toward the Salvation Army canteen.

Until that point, I had latched on to Psalm 23. To me, the crash site was a small picture of the valley of the shadow of death. Without the Lord's promise—"for Thou art with me; Thy rod and Thy staff, they comfort me"—I could never have crossed the yellow tape. I had given that promise to many of the rescue workers, and I saw it bring comfort to them as well.

Bud was demanding more. For him, it wasn't enough to have God with us. He needed the kind of comfort that could only come from a fellow sufferer—someone who had experienced the pain firsthand. These people know it because they've been there. They've felt the grief, walked through it, and come out the other side. Anything less is cheap comfort and doesn't satisfy. For Bud, God didn't fit into this category because He's not a fellow sufferer. He can't identify with us. He's God, all-powerful, all-knowing, not held captive by time and space, and therefore, not afraid of tragedy, dying, or death. He can't understand our pain.

It's not enough that God was with us at the crash site. He's not one of us. He can't identify.

This raises harder questions. Can a married couple, on the verge of divorce, afraid for their children and for their own sanity, turn to a celibate priest for counsel? Can a woman who has been overpowered and raped find solace with a trained male professional who knows nothing of such violence? Can a Vietnam veteran work through the sweats and nightmares thirty years later with someone who has never experienced war? Can a young man at the height of his career soothe the older man who is weary from the long days of retirement with nothing to do—who feels worthless, and tossed away? Can a white person identify with the struggle of the African-American?

For some people, the answer to these questions is yes. They can find comfort with those who know nothing of their problem. But for other people, the answer is no. They need someone who will identify with their pain. Nothing less will suffice. For this latter group, Psalm 23 does not bring the comfort they need. Even those who believe in God and know He is with them in their suffering don't find the promised comfort. Down deep, they remain restless and struggling. They may want to turn to the Lord, but they can't. They feel uniquely separated from Him, alone and physically distanced. The facts are clear and disturbing: The Lord doesn't really know it all. Yes, in His wisdom, He intellectually knows about it. But there's an essential knowledge that's missing. He doesn't know it by experience like they do—firsthand.

Without that intimate knowledge, there's no peace for their restless souls.

For some, this may be the sole reason they don't believe in the Lord. He isn't a help in their time of sorrow. If He speaks with compassion and counsel, wisdom and understanding, it won't matter. It's all cheap comfort until the Almighty can put His face into the face of our pain, our death—and feel it. That's their cry. *That's what Bud meant: Don't let God off the hook. If He cared, He'd be working with us side by side. If He loved us, He'd know what it's like in this hellhole.*

Standing there at the crest of the hill, looking down at the logging road, I knew exactly what Bud meant. That same cry was inside me, too. It was a cry for a Savior who understands.

Christmas at the Impact Site

I stood in my pulpit. It was the midnight Christmas Eve service of 1994, three and a half months after the crash. That night, I was transported back to the Saturday afternoon with Bud at the impact site. I remembered standing at the top of the hill and looking down at the rescue workers as they pulled up the third body. I could still feel the tension of that afternoon. On the one hand, I had just experienced the optimism of Arthur and Sam. They were survivors—just like the perfect soda can—thinking only the best in the midst of the worst.

On the other hand, there was Bud.

The church was full, and the members of the congregation were decked out in all their colorful Christmas garb. I saw the faces of people who never attend church all year round, but they came that night. They were smiling, singing the carols, filled with the joy of the season. Families were sitting together. Kids were home from college. A grandmother who was out of the nursing home held her grown granddaughter's hand. A two-year-old was sleeping in his father's lap. Right next to him, a newborn baby lay quietly under her mother's chin. It was the first Christmas for this new family of four.

It was a special night, a traditional night—a night also celebrated by the world. The world simply replaces the Christian theme of the manger, mother, and Child with the theme of love. Families gather and decorate the homes. Ornaments, dating back to when the children were in elementary school, are put on the tree. Tinsel, outdoor lights, and ribbon around the mailbox complete the decorations. There's the smell of Christmas cookies baking, the sound of Bing Crosby singing "White Christmas," and the sweet joy of opening a holiday card from an old friend. It's the season to buy and wrap presents—bought in love, given in love, and received in love. It's what Christmas is all about in the world. For just a moment, Scrooge sees love as bigger than greed, Tiny Tim as more than his wealth—and love wins.

For many people, the Christmas Eve church service is part of their tradition. They come wanting to sing their favorite carols: "O Come, All Ye Faithful," "O Holy Night," "Away in a Manger," and "Hark! the Herald Angels Sing." The songs, the Bible reading, and the sermon all tell the story of this Child's birth: Joseph and Mary traveling to Bethlehem, the bright star guiding the wise men, the angel of the Lord and the heavenly host appearing to shepherds in a field, an inn with no room for the King to be born—just a simple cattle stall and an empty manger in which to lay the baby. It's a miraculous story, a wondrous night.

It's the preacher's job to keep it miraculous from year to year. People want to feel the joy of Christmas warming their hearts. As far back as I can remember, that warmth always came at the end of the service, with tears and a shiver of goose bumps as we sang "Silent Night." The lights were turned low, and small lit candles were passed out. Together our voices blended, with no accompaniment, until the last verse was sung:

> Silent night, holy night! Son of God, love's pure
> light.
> Radiant beams from Thy holy face,
> With the dawn of redeeming grace,
> Jesus, Lord, at Thy birth, Jesus, Lord, at Thy birth.*

For a second or two, the church would stay completely silent, save the last echo of our voices. It was holy, beautiful,

*"SILENT NIGHT! HOLY NIGHT!" by Joseph Mohr and Franz Gruber. Translated by John F. Young.

and inspiring. Christmas Eve was a night when miracles happened—when God was in His heaven, families were together, children were beaming with delight, and love warmed every heart.

That's what people expected on Christmas Eve, 1994.

Standing in the pulpit, I knew I couldn't give it to them. I had to move beyond the sentimental warmth, the tears, and the goose bumps—beyond the world's message of love and family, and even beyond the church's traditional message of the story of Christ's birth. That night, I had to go deeper, be more real, and tell it like it is. Christmas isn't really about family. It has no frills, no bright lights and tinsel, and nothing to do with giving each other presents of love. The real meaning of Christmas is simply an answer to a prayer—a loud prayer, a painful cry from deep in the soul for God to bring us comfort.

So, I took my listeners back to the impact site, to the top of the hill overlooking the logging road, to that Saturday afternoon. I introduced them to Arthur and Sam. I told them how I had spent all my life like these two men—avoiding death, searching endlessly for the silver lining, finding the best in the midst of the worst. Then I let them meet Bud—the one who forced me to stare at death until I felt it.

The real Christmas message begins with darkness, despair, and death. I wanted the congregation to hear the loud cry in Bud's heart. Without it, the true meaning of Christmas would be lost. So I told them what he said to me as we watched the three bodies coming up from the earth: *God is no help to us. He's not a fellow sufferer. He can't feel our pain, know our suffering, or taste our death. He's all-knowing, but He doesn't know the most basic thing about us—the experience of being human, of living in this fallen, sinful world—this hellhole! He's not one of us, and therefore, He can't bring us comfort in our times of tragedy and death.*

I asked if anyone knew that cry. It comes only from those who have faced suffering and death—faced it and felt it. It's a cry of helplessness that we give with our hands extended—reaching out for God to hear, to come, to know our loss, to give us comfort, to save us from the tyranny of death, and to give us a hope for our future. It's a cry that cannot be satisfied by warm goose bumps or a loving present under the tree.

It's a real cry for a real answer to real suffering from a real God.

I told them that Bud had walked away. In his despair, he had concluded that God couldn't answer his cry. I was sorry for that because God could and He did. And that's what Christmas is all about. It is only for those who know Bud's cry and who, with hands extended and hearts open, are longing to hear His answer. To them and them alone will the message of the Lord's angel to the shepherds make sense: "Do not be afraid; for behold, I bring you good news of a great joy which shall be for all the people; for today in the city of David there has been born for you a Savior, who is Christ the Lord" (Luke 2:10-11). Who will rejoice in this "good news of great joy"? Who needs a Savior but those who are crying out to be saved? For them, Christmas is the answer to their prayers. God in His real love is answering the deepest cry of the soul.

That night in Bethlehem a baby was born. A Jewish prophet who had lived seven hundred years before had told his people about this child's birth: " 'Behold, the virgin shall be with child, and shall bear a Son, and they shall call His name Immanuel,' which translated means, 'God with us' " (Matt. 1:23; see Is. 7:14). He had come. No longer could anyone argue, "God is not a fellow sufferer. He can't feel our pain, know our suffering, or taste our death" because on that night, God came among us. He was born of a human mother. Real God became real man. No longer could anyone say, "God doesn't know us by experience; He can't identify with us." It's not true. He can. He came. He lived. He suffered death. And He lives today, knowing it all. He's one of us. Therefore, He can hear our cry, bring us comfort, and save us in the time of trial.

That's real Christmas. God the Savior is born.

For some, it's hard to understand. They've never had the deep cry in their bellies, so it means nothing to them—this good news that God has come to us in the flesh. Jesus Christ, they say, was a good man, a good teacher, and no different from all the other great religious leaders. For them Christmas remains a family time, full of love. And they come to Christmas Eve services wanting to sing carols, to hear the story of a baby born, and to get that special feeling of warmth.

But that's not who Jesus Christ is at all. Christian history has borne out this one fact: The meaning of Christmas is found in the word *incarnation*, meaning "in flesh." God came to us—*in flesh*. God, who has revealed Himself to be Father, Son and Holy Spirit—one God in three distinct Persons—has entered our history, our time, and our space. Jesus Christ is none other than God the Son. In many churches on Sunday mornings, we Christians recite a summary of our faith. One of these statements, called the Nicene Creed, states what the Bible says about this baby born in Bethlehem:

> We believe in one Lord, Jesus Christ,
> the only Son of God,
> eternally begotten of the Father,
> God from God, Light from Light,
> true God from true God,
> begotten, not made,
> of one Being with the Father.
> Through him all things were made.
> For us and for our salvation
> he came down from heaven:
> by the power of the Holy Spirit
> he became incarnate from the Virgin Mary,
> and was made man.[1]

Incarnate means that God "was made man." He came "for

us and for our salvation."

I wish Bud had been there that Christmas Eve. I have never forgotten him, his cry, or his last request of me: "Don't let God off the hook." How desperately he wanted the Lord to put His face into the face of death—and feel it. I had never seen it that way before, but he was right. And I wish I could have told him that his prayer was answered. God was forever "off the hook." It was true. There was good news of great joy. The Lord had come to the hellhole—come to face it all and to feel it.

And that, I told the congregation, is what Christmas is all about.

---

[1]The Nicene Creed was established in A.D. 325 by the first ecumenical council in the church's history. The purpose of the council was to prevent a schism in the church due to heretical teaching.

Chapter Ten

Death in the Tree

Sunday had come and gone, having left its mark. My strong conviction is that Christians have something to say in tragedy, and this is the subject of the following chapters. Comfort is possible. There is relief on the hard road of grieving, and without question, our church family was grieving. It was sad to see people coming forward at communion with tears in their eyes and pain in their hearts. Erilynne and I wanted to take that pain away, much like parents holding a crying child in their arms, with his knees scratched and elbows bruised. Of course, it's not possible to remove the pain. Comfort is in being there, holding tightly, loving deeply, and speaking clearly.

After the service, all our flight attendants and their families met at the front of the church. There were reports that more than

100 flight attendants from the Pittsburgh area had quit their jobs, effective immediately, all because of the crash. Fear had struck; the accident was too close to home; no one had survived; and no cause had been found. That made the idea of boarding a plane a fearful thing. And how can anyone push fear aside, put on a happy face, and playact for the passengers: "Coffee? Regular or decaf? Milk and sugar?"

I saw that fear in the eyes of some of our flight attendants. One of them had a flight that very afternoon, and her three teenage daughters clung to her arms crying, not wanting to let her go. Would she die? Would they see their mom again? Another woman got calls from coworkers who believed in superstition: "A plane went down in the summer, now this one. Death always happens in threes. There's going to be another crash! We have to quit!" This woman didn't believe in superstition, but fear makes the mind imagine the worst. At times, it's easy to push aside. But other times, like late at night in a hotel in an unfamiliar city, when you're alone and away from the family, it's not easy to dismiss. Doubts come, along with fears, and thoughts of death—of never seeing the kids again—and questions: *What am I doing risking my life? Why not quit?*

"My problem isn't just fear," Johnene Belsky said. "I feel unprotected. Every time I fly, I say my prayers. I ask the Lord to be with me and make the journey safe. I read my Bible in the morning before I leave for the airport and at night when I get to the hotel. The Lord has always answered my prayers. But what about the people aboard Flight 427? There were Christians on that flight! Did you read some of the obituaries? I don't understand. What happened to them? I'll bet they said their prayers. So why weren't they protected? What does that mean for me? Is the Lord going to keep me safe? This crash is confusing me. I feel like I can't trust Him. I know that's not right, but it's how I feel, and I don't know what to do."

Johnene looked at me, wide-eyed, with tears running down

her cheeks, visibly shaken. I knew what she wanted me to say: "Don't worry. The Lord is trustworthy. You will never suffer. Your plane will never crash. You will be protected. Your prayers will always be answered. Don't ever worry." But I couldn't make that promise. For that moment, it was enough to say, "The Lord is trustworthy. He will always be with you no matter what happens. He will not let you go—even in suffering." The other flight attendants consoled her, and together, we had a time of prayer. They asked for the ability to trust the Lord more fully. They asked for His help in putting aside their fears and gaining the courage to leave their families and board their next planes.

As we left the church, I kept hearing Johnene's words ringing in my ears: "I feel unprotected...Is the Lord going to keep me safe?...I feel like I can't trust Him." They stayed with me that night and into the next day. Down deep in my gut, I knew that feeling. She had said more than I wanted to admit. How can we be unsafe with the Lord on our side? How can we not trust Him? But I knew what she meant. That's how it felt. I had expected the worst to be over. It wasn't. The grief marched on, pounding and pounding as it went.

Time to Leave

Sunday afternoon, Ken and I went to the Plaza. We had hoped to go back to the crash site, but by the time we got there, the site was closing early. The members of the media were being allowed to go up and take pictures from a distance. We decided not to stay, instead resolving to meet at the Plaza the next morning.

Monday came, our fourth day at the site. By the end of the day, most of the bodies would be removed. The rest of the week would be devoted to looking for smaller remains, sifting through the dirt with earth-moving machinery. The Delta team had built simple, wooden structures—square, four-legged, and waist high. The tops were made of a heavy wire netting that angled toward a rescue worker who manned each structure. A backhoe lifted the dirt and dumped it onto the netting. The worker's hands sifted the dirt through as the netting caught larger objects. It was exacting work—in a relentless effort to rescue everything.

Ken and I got to the site and decided to split up. By now, many of the workers were known to us by name. We hadn't been back since late Saturday afternoon, so I was anxious to see how people were doing. The fact is, I wasn't doing well myself, and I expected the same from the workers. After four days of seeing the aftermath of violent death, I thought I'd hear more conversation about life's hard questions. It was what I wanted, and needed, but it wasn't what I found. After a few hours, I was uneasy and distraught.

"Did you see the Pittsburgh Steelers play yesterday?" one man asked. Two others around him piped in. Play after play was described in detail and with enthusiasm. I didn't see the game, nor did I care about it. I did my best to playact, exhibiting a degree of interest and then, at an acceptable time, escaping. Another man spent fifteen minutes telling me how he was remodeling his home. A woman needed to talk about her working conditions at a Pittsburgh hospital. She wasn't being treated fairly. Two men spent the better part of half an hour outlining every theory of Flight 427's crash. Both were USAir mechanics. Their shoptalk went over my head. There was no good way to get out of the discussion; I tried and failed three times.

Ken had good conversations with the workers, but not me. I was restless and having a difficult time with all the small talk. I felt the same frustration I often do at a funeral home. There I am, standing in front of the casket of someone I love. People come into the room, pay their respects, and offer their sympathies. The night wears on and people linger. They want to be near, so they come, not knowing what to say and not wanting to feel awkward. They make small talk about the weather, a new recipe, a common friend who is troubled, a story about the children or grandchildren—or worse, they give a bit of advice on how to deal with grief or offer a support group, a book, or a telephone number to ease the pain. Small talk is nothing more than empty words.

I get more restless. I lose perspective. It's often the same conversation I'd be having on a regular day in the supermarket. *But that's not right! We're in a funeral home. Someone I love is dead. My life will never be the same. It's not a regular day. So how can we talk about football—or remodeling—or the weather? How can we talk about common things when life has been flipped upside down and made altogether uncommon?*

That's the frustration I felt that Monday morning at the site. I recognized it and realized I wasn't doing my job. I had come to be with the rescue workers, no matter what they wanted to talk about. But it wasn't working. I didn't want to be with them. I didn't have anything to offer them. I was too caught up in my own grief. I was wrestling inside, struggling, aching for answers to my questions, and feeling wholly dissatisfied until I found them. Late in the morning, I decided it was time for me to leave.

Why Is There Suffering, Lord?

I was struggling on two levels; first, I was struggling with the experience itself; and second, I was struggling with the Lord.

Maybe suffering is easier when it's explained. At age six, I put my hand on our stove's hot burner. I screamed and knew why. But this crash had no explanation. Why was it this plane? Why these people? Why here in these woods? It seemed so random—like a lightning bolt from nowhere. Were these people simply in the wrong place at the wrong time? Don't we have any control over our lives? Does it make a difference if we choose right over wrong—live good, clean, decent lives? Does it matter? If it doesn't, then I've no control of my life. The next accident could happen to me—and that's frightening. I need a controlled, orderly world—not one where suffering is random, without rhyme or reason.

Suffering should be deserved and, therefore, fair. If I can order my life by the right rules, I will know how to live and how to protect myself from suffering. That's fair. That's safety and control. But if suffering is undeserved, I am lost. It doesn't matter what I do. Everyone—the good and the bad—suffer alike. That's chaos! It's unfair! I'm left helpless, unprotected, without meaning to my life, and forced to blindly accept suffering when it comes to me and those I love.

I brought these questions to the Lord, and that made it worse. David Watson said it so well in his last book, *Fear No Evil.* He was an Anglican priest, a gifted evangelist, who was dying of cancer at the age of fifty. The book records the suffering of his last year. At one point, he wrote:

> It is worth noting that suffering becomes a problem only when we accept the existence of a good God...If there is no God, or if God is not good, there is no problem. The universe is nothing more than random choices and meaningless events. There is no fairness, no vindication of right over wrong, no ultimate purposes, no absolute values...If there is no 'good God,' that is the logical consequence, and to protest about suffering is as foolish as to protest about a number thrown by a dice.[1]

The Bible never argues the existence of God. It assumes that we who read its pages have met the Lord. He has come to us and opened our hearts. We believe He is because He has revealed Himself to us. It's the only way I know for certain that God is. He has created our world. The universe isn't spinning randomly, without fairness, justice, or meaning. No, God is the author, and God is in charge.

I believe, as the Bible says, that God is good. Therefore, suffering is a problem worth protesting. For if God is good, why did He allow this tragedy? Why are untold millions left destitute today throughout the Third World? Why are hundreds of thousands of innocent victims dying of starvation and poverty, with no food, no clean water, no doctors, and no medicine to combat their illnesses? Why are there orphaned children who are malnourished, with no homes, no love, and no chance at life? Why has this century been so filled with war, terrorism, and the unjust slaughter of helpless people who've been bombed, shot, beaten, gassed, and brutally hacked. Why did David Watson die at age fifty of cancer? Why are the Pittsburgh police, at the time of this writing, looking for the murderer of an innocent thirteen-year-old girl who was abducted on her way to school, strangled and buried?

If God is the cause of these things, He is not good. If He is all-powerful, why doesn't He stop the suffering? If He is free to intervene, having no limitations, why doesn't He come to our rescue? If He is just, why won't He defend the cry of the helpless? If He is love, why doesn't He take away our pain? If He is consistent, why won't He promise to protect us from harm, *always*? If He is all-knowing, why won't He guide our paths away from suffering before it happens? If He is everywhere, why does it feel as if He is distant sometimes, like a spectator who watches, indifferent and unmoving, unwilling to step in and act?

These were the questions burning in my heart.

In unbearable suffering, it can feel like God is against us. We need more comfort than someone simply saying, "God is love. He is all-powerful, all-knowing, ever present, ever good, always free, always just, always ordered, and always consistent." It doesn't feel like that to the person who is suffering. The emotion is too strong. We reason through our pain that if God were really God, He'd never allow the kind of torture suffering brings. The world would never be in its mess if

God were really God.

In the end, the argument seems convincing: Either God caused the tragedy and is not good, or He couldn't stop it and is not all-powerful. Either way, God is less than God. For these reasons, many turn away from the Lord in times of tragedy. They see no other choice. He will not explain Himself. So they opt for a Godless world in which suffering remains random, meaningless, and unexplained.

Such was the view in *Sleepless in Seattle,* a recent popular movie. In the opening scene, a young boy stands next to his father in a Chicago cemetery, in front of his mother's casket. The child wants to know why she died. The father does his best to answer, but even he doesn't know why his beloved wife died in the prime of her life. Gently, he says to his son: "Mommy got sick. And it happened just like that. There's nothing anybody could do. It isn't fair. There's no reason. But if we start asking 'Why?' we'll go crazy."[2]

This is the world apart from God; it's unfair, there's often no reason why, and it's better not to ask.

Another choice people make is to turn to the Lord with passive acceptance and resignation. These people accept suffering as an inevitable part of life. It happens to everyone. We must bear it, come what may. Moreover, they say, we must stop questioning the Lord. After all, our minds are too small to understand the whys of life or the ways of God. Some things remain a mystery; accept them and resign the pursuit, lest our lives end in frustration and bitterness. We must trust that God is sovereign. He loves us. He knows what's best for us.

Again, it's better not to ask.

I could not do that, I couldn't blindly accept undeserved suffering. It does not satisfy the broken heart, nor does it tell me what to say to Johnene at church who feels unprotected. Even with the Lord God on her side, she doesn't feel safe. She's afraid to trust Him, and she doesn't understand why. She's a mom. She would never allow harm to touch her three-year-old child. Never! So why can't she have the same confidence in the Lord? Why can't she feel safe? Must we avoid asking the Lord why? Won't He answer? Why won't He bring us comfort?

That's what I was searching for late Monday morning at the crash site. There had to be answers. God is God. He has revealed Himself in the Bible. He is good. He is all-powerful. He is holy and just. His character is beyond reproach. Jesus said, "Ask, and it shall be given to you; seek, and you shall find; knock, and it shall be opened to you" (Matt. 7:7). I was asking, seeking, and knocking. There had to be answers.

Standing on the logging road just below the impact site, I lifted my eyes into the deep blue September sky and prayed silently, "Father in heaven, help me to understand. I can't believe that You have left us suffering without answers, without hope. I need to know. In Jesus' name. Amen."

The Tree

I spotted Ken down the road, talking to a worker. I walked toward him, hoping to catch his eye and signal that I was headed to the canteen for a break. But before I got to him, I saw the most beautiful black Labrador retriever coming up the road from the tent. He hugged his master's side, intently obeying his every word. When his master stopped to talk, he sat down at once. I watched him panting away, waiting patiently. I couldn't help drawing close, hoping his master would finish the conversation.

"May I?" I asked when he was done, wanting to meet his dog.

"Sure. His name is Brier, since that's what's all over him after a day's hunt." The man looked like he was in his late sixties. He had a rough, commanding presence. He was a no-nonsense type.

"You've trained him well," I said, stooping down to pat an excited Brier.

"He's a good dog," the man responded in his low, husky voice. "He's about four. One of my best. Some dogs train faster than others. Brier was a fast learner. Picked up my commands in no time."

"How many dogs do you have?"

"Right now, we have about eight. But I've been training black labs for police work some forty years—mostly narcotics, drugs. But this here's a cadaver dog. I've trained him to find the dead."

"A cadaver dog?"

"I've been on the police force all my life. I used to train dogs when I was off duty, but now I'm retired. I've got more time. Precincts from all over western Pennsylvania get me to train their dogs. Nothing can beat their noses. They're good dogs. They do good work. Now, Brier is mine. He even stays in the house with me and the wife. Fine dog. I trained his pap and his grandpap, back five generations."

"But how do you train a dog to find the dead?" I asked. I didn't get it.

"The police gave me a guy named Fred. So I trained Brier on Fred." Just the mention of Fred made Brier stand up, stop panting, and look directly at his master. He was ready to go. "I tell the dog to go find Fred. When he does, he barks and then sits. He won't go anywhere until I get there." He patted the dog's head, ordering him to sit. He smiled with pride, knowing Brier was ready to work.

"We're here to find the rest of the remains. Huh, boy?" the man said.

There was work to do. Brier wasn't here to play or bring enjoyment. What an odd twist for me. My dogs have always been a wonderful distraction after a stressful day of work. Our Old English sheepdogs greet me at the door, wagging their rear ends, and barking with excitement. Then it's time for a bone, and a walk down the street—a chance for me to clear my head, get a new wind, and see things from another perspective.

But Brier wasn't here to distract us from our work. He'd shoot through the woods, bark, sit, and make us face death all over again. He was a cadaver dog, eagerly waiting for the moment of release.

"You want to see him work?" the man asked me. I nodded with interest.

"Okay, then," he said to Brier, unleashing him. "Go find Fred!" Off Brier went with a yelp, into the woods opposite the impact site, and out of sight. I decided to follow the man into the woods after Brier. I wanted to see the dog in action. But the moment Brier disappeared, the man was interrupted by a police officer. They began to talk, and I could tell it was going to be a long discussion. I quietly backed off, crossed the road, and slipped into the woods after Brier, wondering if I'd catch up to him.

At first, I didn't see him. I went deeper into the woods, alone. I surprised myself. I was supposed to be leaving the site. All my questions, all my struggling with the Lord—and I was chasing through the woods after a cadaver dog! It didn't make sense. Just then, I heard Brier bark from up ahead and to my right. I made my way there and, in a short time, saw the dog there sitting at attention, his eyes fixed, still barking.

I came up behind him. I noticed that the area was somewhat opened. There were fewer trees, so more sun was coming through. I didn't touch Brier; he was too intense. He had smelled the dead. But I couldn't see it. I came up next to him and studied the ground in front of him. Nothing. Five feet in front of him was a tree—a young tree, only twenty feet in the air. Most of its leaves were gone, maybe from the fire. I kept looking at the ground, searching. I couldn't see any remains. But Brier kept barking—and staring.

He was staring up.

So I lifted my eyes up into the tree, and there I saw the dead.

A chill went through my body. I stood there in silence, staring. I was suddenly, surprisingly aware that this was it. The

Lord had heard my prayer. He knew my wrestling. He had brought me here to see the answer to all my questions. There, in the tree, was a body suspended between heaven and earth—a body beyond recognition. I looked carefully, but I couldn't identify it. The remains were too disfigured. But I knew it was human from the bones, the raw flesh, and the burnt blue jeans. The body was pressed against the bark of the tree, hanging. The sight was repulsive—all too vulgar.

All too holy.

Ken came up behind me. He saw it and froze. There we were, both of us, looking for a moment in time through a window into the heart of God. The body in the tree was no different from our Lord's body, pressed hard against the bark of the cross. His body had been broken, beaten, bruised, tortured, whipped, nailed, pierced, and left hanging—suspended between heaven and earth, abused beyond recognition. The Bible says, "His appearance was so disfigured beyond that of any man and his form marred beyond human likeness" (Is. 52:14, NIV). His bones, His raw flesh, and His dead body without human likeness were just like this one. We stood there in silence as if we'd both been taken back in time—back two thousand years to that place outside the gates of Jerusalem. It was as if we'd been given permission to stand before our Lord's cross and to see His dead body—to catch a glimpse of His suffering.

God had come among us in the flesh, knowing our pain—on a tree, knowing our death.

"Look, Ken," I said in a whisper, breaking the silence. "It's like the cross of our Lord and Savior, Jesus Christ."

"I can't help thinking," Ken said, "of that passage in the Bible which says, 'He was despised and rejected by men, a man of sorrows, and familiar with suffering. Like one from whom men hide their faces he was despised, and we esteemed him not' " (Is. 53:3, NIV). That was the Lord Jesus.

I wanted to stay there with the window of God still opened for just a while longer.

But it didn't last. Brier had summoned his master. We heard the sound of men coming toward us. His master came first. He praised the dog, gave him a quick pat, and a treat from his pocket, and put him back on the leash. Another man said he'd get a coroner's team and a ladder. Ken and I decided to wait. A few minutes later, a coroner came, set the length of his ladder and hoisted it up the tree, resting it only a few inches from the body. Then he climbed up and carefully examined the remains and called out the medical terms, naming each body part. A man next to us wrote down his words. A photographer got his camera ready.

With precision, the coroner separated the impaled body from the bark of the tree. Then slowly, he brought it down the ladder rung by rung. A team member waited at the bottom with an open body bag. While it was still suspended in the air, the photographer took pictures of the body from different angles. When he was done, the remains were placed in the body bag, labeled, and sealed. Down came the ladder. The body was taken to the refrigeration truck. The coroner team left. And Brier was told to go find Fred.

The window closed, but it had been open long enough. The Lord had heard my cry—and answered.

Perspective

Standing at the tree, I knew: God has not kept His silence. He has spoken. He has acted.

My perspective changed. I stopped looking at God through my own pain, and in the context of my small, self-centered world. Instead, I started to see Him as acquainted with suffering, hanging on that tree, knowing death.

God is good. The Bible never accuses God of authoring evil. It says, rather, that He created the world in perfection and made man and woman in His image. There was no sin, no suffering, and no death. But we rebelled. We chose to disobey the Lord and His ways. It was our doing—our choice and not His—that sin came into the world and marred the perfect creation. With sin came suffering, evil, and death. Our world became fallen, just like us. It was God's plan that we live in communion with Him, but because of our wrong actions, we lost that relationship. Our sin separated us from Him and His intended plan for us.

Through the centuries of church history, and in Jewish tradition, the truth of the Bible has been upheld: The Lord is good, holy, just, and pure. It is wrong to blame Him for the suffering in the world today. He didn't make the plane fall from the sky. He didn't kill the 132 people aboard, nor does He make the cancer and disease that attack our bodies. He doesn't cause famine and poverty to starve millions in the Third World. He doesn't start wars or desire that the innocent and helpless suffer injustice. He isn't the author of fires, floods, drive-by shootings, drunk drivers, husbands who batter wives, or drug pushers who murder teenagers, or anything that is meant for evil and brings us suffering.

God is love. Even though we turned away, the Lord did not abandon us. I experienced His love when I first came to the crash site. I knew His promise in Psalm 23 to be with me "in the valley of the shadow of death." That promise gave me comfort. He doesn't stand outside our suffering and leave us as orphans. Again I saw His love in Bud, the rescue worker. That man needed a deeper comfort. He needed God to come and be one of us. Only then could God understand our suffering. He had to see what we see, feel what we feel, firsthand. To Bud, that would be comfort. And God did just that. He came—fully man and fully God, born of the Virgin Mary—into our history as one of us. His love met us face to face.

But He showed His love in an even greater way. I saw it when I stood there in front of the tree.

God is our Savior. Just the coming of Jesus Christ, the Son of God, to earth to experience our suffering doesn't answer our deepest problem. We remain in a fallen world, with suffering, evil, and death all around us. Whether we want to admit it or not, we also remain sinners. We still rebel against God and His ways. We are born fallen—bent on doing wrong. The psalmist David wrote, "Surely I was sinful at birth, sinful from the time my mother conceived me" (Ps. 51:5, NIV). Our deepest problem is this: Our sin still separates us from having a relationship with the Lord. Our only hope is for our sin to be dealt with and forgiven.

I met a lot of people during the days after the crash who asked, "Why did God do this? After all, God had the power to prevent this accident, didn't He?" The answer is yes; He is all-powerful. And yes, He has the power to stop all human suffering. But suffering came into the world second, not first. What came first was our rebellion—our choice to turn away from God. With all His power, He can make this world new again. But what He can't do—by His own choice—is overpower our right to choose.

First things first: Jesus Christ came to forgive our sins. He came to save us and give us choice again—the choice to follow Him. It is our choice to be in a right relationship with Him.

God is Sufferer. God's plan to save us was both costly and loving. His own justice required the fair, legal punishment for our sin. That just punishment, the Bible says, is death (see Gen. 2:16-17 and Rom. 6:23). But His mercy provided the only possible solution. He would take that punishment Himself. He would take our sin. In fact, He would take all sin for all time. God the Father acted by laying upon His Son "the iniquity of us all."

> We all, like sheep, have gone astray, each of us has turned to his own way; and the Lord has laid on him the iniquity of us all (Is. 53:6, NIV).

The cross is everything! On that day, Jesus Christ received

to Himself all our sin. So instead of us dying for our rebellion as justice demanded, He did it for us. He died our death. Nails "pierced" His flesh:

> But he was pierced for our transgressions, he was crushed for our iniquities; the punishment that brought us peace was upon him, and by his wounds we are healed (Is. 53:5, NIV).

At the cross of Calvary, the Son of God experienced suffering firsthand—not just the horrible suffering of crucifixion, but all the suffering, all the evil, and all the death in the world. On that day, the Lord did more than observe our suffering. He entered into it, once and for all. He took our sin, bore our punishment, and endured our death. God has acted on our behalf, and because of it, "we are healed."

God is life. He has given us choice back. When Jesus Christ rose from the dead, He told His followers to go and proclaim good news: We can choose to turn from our wrongdoing and receive His love, the forgiveness of sin. We can enter into a relationship with Him again, for God's love sent His Son into the world so that "whoever believes in Him should not perish, but have eternal life" (John 3:16).

The choice is now ours. We can believe and have eternal life, or we can continue to go our own way and "perish." While Jesus suffered on the cross, the man next to Him believed. He was a thief and deserved to die for his wrongs. But in his suffering, he cried out to Jesus Christ, and Jesus heard him and saved him. Jesus said, "Truly I say to you, today you shall be with Me in Paradise" (Luke 23:43).

To be honest, most of us don't want to hear that. We pray for the Lord to do one thing and one thing only: stop our suffering, free us from our pain, and give us more time on earth. It's all we want. It's all we see from our small, self-centered world. But the Lord Jesus Christ has something far greater to give. He gives life—eternal life, with no more death! With His words, "You shall be with Me," He promises a forever relationship with Him. And He promises paradise, which is nothing less than a new heaven and earth, created in perfection and without suffering. There will be no more mourning or crying or pain—and no more death (see Rev. 21:3-4 and 7:17).

God has answered! He has not kept His silence! He has spoken. He has acted.

The Lord never said we'd be free from suffering in this world. Until paradise comes, the world remains as it is, with sin, suffering, evil, and death. There are times I wish it weren't true. I desperately wanted to comfort Johnene at church by promising her, "You will never suffer! Your plane will never crash!" But that wouldn't have been true. This is a fallen world. Suffering will always be part of this life.

But something far better has come. We can choose to receive God's love in Jesus Christ and live with confidence and courage. Tragedy will happen. Death will steal those we love and, in the end, lay our bodies in the dust. But no matter what happens to us, Jesus Christ promised, "I am with you always, even to the end of the age" (Matt. 28:20). He will be with us through it all. And what if we board a plane and it goes down in a ball of fire and smoke? What if we pick up the phone and hear that death has taken our child, fire has burned our home, or disease has come to our body? When suffering comes, and it will, we can know for certain that the Lord has not and will not abandon us. We will not perish. He will be with us, even if it comes suddenly, in twenty-three seconds. For at that moment we'll be with Him forever, in paradise.

The choice to believe is now—before suffering comes, before death comes, before it's too late. The choice is only possible because of Jesus Christ—because His body hung on a tree two thousand years ago.

Solidarity

"I could never myself believe in God, if it were not for the cross,"[3] wrote John Stott, an Anglican priest and theologian. Without the cross, God is separate from our human suffering. There is no solidarity. But in the cross of Calvary alone, God in Christ has united Himself to our experience. He is able to meet us in our suffering—no matter what it is—and bring us true comfort and real hope.

He knows suffering fully, intimately.

What happened on Calvary is understood by those who have cried out to God in suffering. They have wrestled with Him, demanded that He answer their questions, and begged for Him to comfort their pained and anguished souls. For these people, the cross is everything. It is God's saving love to their hearts. But some have never cried out, wrestled, demanded, or begged. For them, what God did on the cross in Jesus Christ has little meaning. It is not an answer for people who aren't asking the question.

John Stott, in his playlet titled *The Long Silence,* lets the cry of the suffering be heard loud and clear. And with this cry, there is given the blessed privilege to see and hear the Lord's response:

> At the end of time, billions of people were scattered on a great plain before God's throne.
>
> Most shrank back from the brilliant light before them. But some groups near the front talked heatedly—not with cringing shame, but with belligerence.
>
> "Can God judge us? How can he know about suffering?" snapped a pert young brunette. She ripped open a sleeve to reveal a tattooed number from a Nazi concentration camp. "We endured terror...beatings... torture...death!"
>
> In another group a Negro boy lowered his collar.

"What about this?" he demanded, showing an ugly rope burn. "Lynched...for no crime but being black!"

In another crowd, a pregnant schoolgirl with sullen eyes. "Why should I suffer?" she murmured, "It wasn't my fault."

Far out across the plain there were hundreds of such groups. Each had a complaint against God for the evil and suffering he permitted in his world. How lucky God was to live in heaven where all was sweetness and light, where there was no weeping or fear, no hunger or hatred. What did God know of all that man had been forced to endure in this world? For God leads a pretty sheltered life, they said.

So each of these groups sent forth their leader, chosen because he had suffered the most. A Jew, a Negro, a person from Hiroshima, a horribly deformed arthritic, a thalidomide child. In the centre of the plain they consulted with each other. At last they were ready to present their case. It was rather clever.

Before God could be qualified to be their judge, he must endure what they had endured. Their decision was that God should be sentenced to live on earth—as a man!

"Let him be born a Jew. Let the legitimacy of his birth be doubted. Give him a work so difficult that even his family will think him out of his mind when he tries to do it. Let him be betrayed by his closest friends. Let him face false charges, be tried by a prejudiced jury and convicted by a cowardly judge. Let him be tortured.

"At the last, let him see what it means to be terribly alone. Then let him die. Let him die so that there can be no doubt that he died. Let there be a great host of witnesses to verify it."

As each leader announced his portion of the sentence, loud murmurs of approval went up from the throng of people assembled.

And when the last had finished pronouncing sentence, there was a long silence. No-one uttered another word. No-one moved. For suddenly all knew that God had already served his sentence.[4]

---

[1]David Watson, *Fear No Evil* (Wheaton, Ill.: Harold Shaw Publishers, 1984), 111-112.

[2]*Sleepless in Seattle*, (TriStar Pictures, 1993).

[3]John R. W. Stott, *The Cross of Christ* (Downers Grove, Ill.: InterVarsity Press, 1986), 335.

[4]Ibid., 336-337.

Part IV

Proclaiming Good News

Chapter Eleven

Something to Say

In December of 1944, Betsie ten Boom was dying in Ravensbruck, a Nazi extermination camp. She was sick with fever, her body wasted to bones, and her strength gone. She and her sister Corrie were forced to experience the violence of Nazi Germany—the overcrowded prisons where fourteen hundred women were packed like animals into a room made to sleep four hundred, where so many experienced hunger, filth, fleas, lice, nakedness, cold, disease, slave labor, whippings, and death. Day and night, they heard the screams of women being beaten; they saw dead bodies lined up in rows; and always, they smelled the smoke of death from the crematorium.

There, in a place close to hell, surrounded by suffering beyond human description, Betsie and Corrie found the Lord to be their only source of comfort and strength. In Corrie's book, *The Hiding Place,* she tells the miraculous story of smuggling a Bible into the camp.[1] In Barracks 28, the sisters held worship services where that Bible was read and taught, where prayers were spoken, and where they watched "women who had lost everything grow rich in hope."[2] One service grew into two, without the guards seeing. Women crowded close to hear the saving message, the only good news they could find in the camp.

As Betsie lay dying, she knew her sister's mission in life. It was Corrie's duty to tell the world the secret they had found in spite of the gruesome pain of their circumstances. People had to know what they had experienced. Before her early death in Ravensbruck, Betsie whispered these words to Corrie, "We must tell people what we have learned here. We must tell them that there is no pit so deep that He is not deeper still. They will listen to us, Corrie, because we have been here."[3]

Corrie survived Ravensbruck to fulfill that mission. She could speak because she had been there. She knew pointless suffering—the depth of the pit—by experience. That firsthand knowledge fills words with substance. People will listen to us because we have been there. And the opposite is true: People will not listen when we have not been there. Our words become empty promises with no depth, no life, and no hope for the grieving. The bottom line is that sufferers need words of comfort from fellow-sufferers.

Corrie ten Boom knew the pit, and she knew that the Lord had met her there. Every Friday, the women of Ravensbruck stood in line at attention, their naked bodies emaciated, "unloved and uncared for." They were humiliated and laughed at by a "phalanx of grinning guards." But there she remembered her Lord. He knew this humiliation. He was jeered as He hung naked on the cross.[4] Corrie met Jesus Christ in all her sufferings. She knew that "there is no pit so deep that He is not deeper still." So she traveled the world telling people what she and her sister Betsie had learned in Ravensbruck, continuing until her death in April of 1983 at age ninety-one.

"He is deeper still." Corrie had something to say.

The Beginning of Words

I entered ministry in a time when many pastors were in a period of transition. The old model of pastoral ministry was cold and uncomforting. The pastor was there to perform services, and that was it—no personal involvement with the people. If he came to the hospital or to the funeral home to see a family, his visit was brief. He'd say the right prayers and be gone. The funeral service he gave would be the same one he gave when the last person died—same words, same sermon, different family, different body.

My instructors rejected this model. In its place, they taught us the opposite: Be there! Stay with the grieving families in the hospitals and funeral homes. Get personally involved.

At one point in my training, I became a chaplain at a state mental institution. There I stayed, on the ward for the severely impaired. Words didn't work there, for the patients didn't understand them. So for eleven weeks, I learned to "be there," on the ward, and in the dining hall, present and quiet. Until that time, I had never seen such human suffering, nor had I felt so helpless. There was nothing I could do, nothing I could say to comfort them in their illness. But that was the lesson I was to learn. Comfort, my teachers said, is the act of staying with the suffering. It is wrong to see people in pain and walk away, hard and heartless. Christian love must get involved.

I agreed. I'd never have gone to the crash site otherwise. But comfort requires more than just "being there." For example, a patient suffering in an emergency room finds comfort in the doctor's presence. But that presence isn't enough. The patient needs the doctor to act. The same is true with those who are grief stricken. People need more than a pastor's presence. They have questions, concerns, doubts, and fears.

They must have words.

The moment my phone rings and I learn that somebody has died, I know that in two or three days' time, I will stand before the grieving and speak. The funeral service starts. We sing our songs, pray our prayers, and then I go to the pulpit. I must have words. The act of burying the dead demands it. Mourners come looking for comfort. They want their deceased to be honored, and their lives to be remembered and appreciated. Most of all, the sufferer needs to hear something of God. Where is He? What does He have to say?

The deeper the suffering, the more people demand the right words of comfort. When a man in his late twenties dropped dead of a heart attack, I was called upon to speak at his funeral. In his life, he had been an atheist. He mocked his wife for attending church, for praying, and for reading her Bible. He cursed God openly in front of his friends. Now suddenly, tragically, he was dead. Family and friends packed out the small chapel, which was jammed with flowers; they waited for me to stand up and speak. Everyone was in shock as they stared at the body in the open casket. The widow was crying audibly, with her children at her side.

She needed the right words. She needed someone to make sense of her tragedy—someone who could help bring order to her life, who understood her suffering and knew the secret to her pain. *The secret*—that's what she was looking for! The man's friends stood up to remember him. They told moving stories through tears of love. The man was honored. But the widow and many others needed a deeper comfort, a lasting comfort. They needed to know where God was in their tragedy. Did He care? Did He love?

They need to know that there is no pit too deep.

It's the same as what happened after the crash of Flight 427. I had to speak to the media when they asked on live television, "Tell me, why did this happen?" I had to speak to our church family from the pulpit the next Sunday morning. Both ABC and CBS came with their cameras rolling, waiting for me to give them a sound bite for the evening news. I had to speak to the rescue workers at the crash site and down at the Plaza. I had to speak to the families and friends of the victims at the memorial services. I had to speak to local members of our community who called on the telephone or stopped at the church. They needed answers. That's the demand, deep in the soul, in a time of terrible tragedy. It's a cry for true comfort.

They needed words—words from someone who knew deep suffering, who had asked the same questions and found the secret. "They will listen to us," Betsie said confidently, "because we have been here." These "who have been there" speak words of substance! They are filled with God, His love, His suffering, and His promises.

To tell His secret—that He is deeper still—is for us the beginning of our words.

The Blundering Pastor

"Tell me, why did this happen?" At the moment that question is asked, the pastor stands in the spotlight. Something has to be said. But here's the question: *Do we have something to say?* I have watched the evening news cover many tragedies where reporters ask local pastors the question "Why?" Too many times, these pastors have no clear answer. They stumble, hemming and hawing. They're confused and unclear. Some give the impression that the Lord is just as surprised by the tragedy as we are—as if He knew nothing about it beforehand. Others wax eloquent, becoming philosophical, but in the end they say nothing. Others cry in resignation, "Who knows? It's a mystery we must accept. God has His reasons." Still others say that God didn't do it; it was the Devil. Or they say that God did it as a punishment for sin. Still worse, some pastors blubber saying they love God, but now that He made people suffer and die, they don't know anymore. The blundering pastor, back and forth.

We stand in the spotlight, and people wait to hear.

I have found the same to be true at funerals. Some years ago an acquaintance died suddenly. He was a prominent lawyer in our town, a large man in his late forties, who had boasted that he had not seen his doctor in years. John was a family man, a quick-witted practical joker who had a heart for underprivileged children. He loved his job, his friends, and his summers on Cape Cod. One day in the prime of his life, he had a massive stroke and died instantly, as if someone had shot him. None of us could believe it.

The church was packed; it was a huge, standing-room-only crowd. There must have been a thousand people attending the service. Erilynne and I arrived early and saw many of John's close friends. Everyone was in shock. We all had a recent memory of John when he was healthy, joking, and full of plans for the next week, next month, or next summer. One man had stopped at his mailbox before coming to the church and found a letter from John. It was postmarked the same day he died. This man hadn't opened the letter yet. He couldn't. It was still too hard. How could John be gone? How could life march on without him? It didn't make sense. And there this man was, dressed up, waiting for John's funeral to start; it was unimaginable. It wasn't right, wasn't fair.

There was no casket; the body had been cremated. After a few hymns, prayers, and readings from the Bible, the preacher stepped into the pulpit. More than anything, he needed to help the congregation with the shock of John's sudden and early death. John's widow, three children, and close family and friends especially needed to know the Lord's comfort and love. The chosen preacher had been friends with John since college. Surely he could bring the right words, the necessary words. He had experienced this loss personally.

But it didn't happen.

The preacher told a collection of funny stories—sketches from their life together over many years. It was nothing more than a long toast, honoring John with playful jest. It felt more like a retirement party than a funeral. In fact, the preacher was so effective that I looked around for John. Where was he? Was he laughing? I had to remind myself that he wasn't there. He was dead. On the stories went, each one denying the fact that John was gone. It seemed that even the preacher couldn't face it. At the end of his talk, his closing words made the loss all the more difficult. "John," he said as if John were there sitting in the pew next to his wife, "we've had a lot of great times, haven't we?" And that was all. He stepped down with a smile on his face.

To me, it was a cruel thing to do—as cruel as sending away the hungry without food, the cold and poor without a coat, or the sick without medication. The grieving were not comforted. John had died, and all of us knew it—except the preacher. A week before his same words would have made us silly with laughter. But now they were like salt on an open wound. We needed to face John's death—not avoid it. We needed to cry, grieve together, and seek the Lord and His mercy in our time of suffering. We needed to hear from Him and receive His love. But this was denied us. We left the funeral with no closure, no comfort in our grief.

Denial comes in many forms, but it always has the same result: *We don't face the fact of death.*

One minister, at a memorial service for a victim of the plane crash, told the congregation not to grieve. He said that the dead woman was still here—alive! We couldn't see her, but she could see us. We could talk to her, but she couldn't talk back to us. She wasn't dead. In fact, he said, it's better! "She is closer to us than ever before. And wherever we go, she will be with us." With that, he said a prayer to her.

The dead are not dead? The relationship is better? No reason to grieve? The act of talking to the deceased is found in the black magic of the occult, with mediums and séances in which the living conjure up the dead. In Asia and Africa, ancestral worship is common. The living believe that their dead relatives are still on earth, living as invisible spirits with the power to either bless or curse their lives. In America, comic strips and dime store novels portray dead relatives as helping their offspring through tough times. It may bring cheap comfort to some for a while, but it denies the fact that the dead are dead. Our relationship with someone who has died has been severed. It's not better. Our loss demands that we grieve for those we love and weep for those we miss.

Moreover, the Bible condemns the practice of talking to the dead. Jesus Christ made it clear in His teaching that when we die, we do not stay on earth. In one story He told, a rich man died and ended up in Hades—the place of torment. There, the rich man cried out for someone to come and "dip the tip of his finger in water and cool off my tongue; for I am in agony in this flame." Another man died—a poor man—and "he was carried away by the angels to Abraham's bosom (see Luke 16:19-31)." One man was in hell; another was in the glory of God. Neither was on earth. In fact, the rich man wanted to go back to earth and warn his family lest they end up in Hades too. But God didn't allow it. The dead do not interfere with the living.

That's the fact, and the Bible makes it plain. But the minister had nothing more to say—no words of true comfort. It was easier to pretend that the woman wasn't dead but alive, there in the church, watching over her relatives. Some people may have found comfort in his words, but it was false comfort. It denied everyone the right to face her death, to grieve their loss, and to turn to the Lord for solace in sorrow.

In another service, a pastor actually blamed the dead man for flying on Flight 427! The man was scheduled for a later flight and "in his unwillingness to wait" brought the accident on himself. Who can believe that logic? But it's not uncommon these days. I've heard other pastors blame the dead for drinking, smoking, driving too fast, and not following their doctors' orders. "Had he acted right" the logic cries, "he wouldn't have died, and no one would be grieving." Again, these words are said so that we don't have to face death. Blaming anyone for our pain does not bring the dead back, nor does it bring deep comfort to the soul. It's wrong to do and a real blunder.

Here's one more story. At another memorial service for a Flight 427 victim, I noticed that the preacher was terribly nervous. He wanted to comfort the people by facing death head-on, but he didn't know how to do it. So he recreated the flight. With his words, he made us live through those fatal twenty-three seconds. It was as if we were all on board the aircraft. Then he asked the most terrifying question of all: "What were those last twenty-three seconds like?" The more he talked, the more the twenty-three seconds felt like an eternity. "Did your loved one cry out for God in those last seconds?" he asked. Maybe, just maybe in the sheer panic, the person's last thoughts were on the Lord. And that, the preacher hoped, would bring dramatic comfort to the families.

It didn't. It was painful to live through his recreation of the flight and to wonder what the people endured at the end. Did they cry out to the Lord? Maybe—and "maybe" doesn't give anyone comfort.

Where are the words—the right words? People deep in the pit of suffering are waiting to hear.

The Need to Hear

"I should have called you earlier," Smitty said. We sat in a fast-food restaurant near the airport in the mid-afternoon. Smitty had called me a month after the crash and wanted to meet. He was a USAir mechanic who had worked that week at the site. His job was to make sure all the plane parts were properly marked, decontaminated, and loaded on the trucks. I recognized him, but we hadn't met at the site.

"I saw you there," Smitty said, "working with the coroner's teams. Seeing chaplains at the site made such a difference. It made me realize that the Lord was there. He was with me. He'd see me through."

"Smitty, I could never have gone to that crash site without Him," I said. "I'm a priest. I see death all the time. But I've never experienced suffering like that before—nothing like that."

"What I don't understand," he began, "is why I did so well during the week. Don't misunderstand me. I hate death. I hate funeral homes. I stay away from them. Seeing the dead at the site was the hardest thing I've ever done in my life. At times it got so bad, I had to walk off the site just to clear my head, you know? Sometimes, I cried when nobody was around. Other times, I prayed to the Lord. I needed His strength to go back. And I always got it. I went back every time. I stayed at the site the entire week."

"And you made it through all right."

"Yes, I did. And I'd go home at night, eat a full meal, and sleep like a baby. I'd wake up the next morning and head back to the site. That happened all week. I knew I was supposed to be there." For an hour, we talked about our experiences at the site—things we saw, people we met, and the suffering we felt. It was good to tell the story again to somebody who knew it—who had been there. I found I really needed it.

"What has it been like since?" I asked later. I knew something was bothering him. Both of his oil-stained hands fiddled with the napkin, then with his coffee cup. His eyes kept brimming with tears. I guessed that Smitty was in his early fifties. He had a medium build, a bushy dark brown mustache, gold wire-rimmed glasses, and a full head of graying brown hair that was neatly combed and in place. He told me he had one child at home and another who was grown and on her own. He and his wife were devout Catholics, faithful to their church.

"Well, that's the problem. I went back to work the next Monday morning, put in a full day, went home, had dinner with my wife, watched some TV, and went to bed. That's when it all started. I couldn't sleep. I kept seeing dead bodies everywhere, like I was back at the crash site. I started getting afraid—"

His voice broke and he put his head down, as if he was embarrassed. He took his glasses off, rubbed his eyes with the napkin, and went on, "—of dying. I'm afraid to die—how I'll die and where I'll go. It's been a month, and I still can't sleep. In the middle of the night, I get this insatiable desire to eat. At four o'clock in the morning, I'm having another full meal! I've never done that before. And I cry at the drop of a hat. My wife told me to see our family doctor. He sent me to a shrink who gave me pills and told me I had post-traumatic stress syndrome. The pills are supposed to regulate my moods and help me sleep."

"Have they worked?"

"Not really. I'm sleeping some, but I still can't shake the sights of what we saw up there. And I can't stop thinking about dying. I wake up with sweats like I've worked out at the gym. It's affecting my job, social life, home life—you name it. When is it going to end? I can't keep taking these stupid pills all my life. I decided to call my priest. He's been a good friend for many years. He listened to everything I said—real sympathetic. He prayed a nice prayer for me. But deep down I knew he didn't understand."

"He wasn't there, Smitty. Sometimes we need people who know our suffering."

"That's why I called you."

"Well, you're asking the right questions. We have to face death," I said. "I remember the Sunday afternoon memorial service on September 25 at the Plaza. When it was over, the families were taken up to the crash site by bus. It was the first time they were allowed to walk down that old logging road and see for themselves where the plane had crashed and where their loved ones died. I'll never forget it."

"I heard about that."

"Well, let me tell you what happened. My bishop and I walked up the hill toward the impact site and turned around. When everyone became quiet, the bishop opened with prayer. Then I described the crash site for them. At the end, I pointed to the cross behind me made of white carnations with a strand of red roses across its front. It was placed, I told them, at the actual point of impact—where the nose of the plane hit the ground. Well, when I was done, the most amazing thing happened. None of us expected it. We thought the people would stay for a few minutes and then go back to the buses. But guess what! They climbed the hill! Young and old, hundreds of people came. They were driven to stand at the impact site, to touch the trees, walk in the woods, feel the dirt in their hands, look up into the sky past the tops of the burnt trees, and see for themselves. They physically needed to face the death of their loved ones."

"I can't imagine what they've lived through," Smitty said with sadness.

"Nor can I. And it was harder for them. They didn't have a body to see—nothing to touch or hold. That's why being at the crash site was so important. It was their one chance to get as close as possible. One woman in her seventies, I remember so well. She took a branch and scraped the dirt, digging and digging like she wanted something—maybe a piece of the plane, something she could hold in her hands and take home with her. See, she needed to face her loved one's death—and grieve."

"I can't face death," Smitty recognized. "I'm afraid of what's going to happen to me."

"Well, what is going to happen to you?" I pushed. He shook his head, he didn't know. "During the week at the crash site, you asked the Lord to give you strength. He did. He answered your prayers. You knew He was with you. Well, the same is true when we die. He isn't going to abandon us."

"I don't understand. You've got to tell me what's going to happen to me." Smitty leaned forward, ready to listen, waiting

...ar. He was pleading with me. The pictures of death in his memory from the crash site were too real, too overwhelming. He wanted to face death—his own death. He didn't want to spend the rest of his life taking pills, overeating, and not sleeping. He needed answers. He needed them now. And he needed to hear them from someone who had been there and would tell him the plain facts—nothing more and nothing less.

"Okay. Let's start with communion. Why do you take bread and wine on Sundays?" I asked.

"Because Jesus Christ died?" Smitty answered with hesitation.

"That's it. He told us to take the bread and eat. It represents His body, broken for us on the cross. And He told us to take the wine and drink. It represents His blood shed for us for the forgiveness of our sin. Every Sunday, you face death at communion—*His death.* The bread and wine point to what Jesus Christ did when He suffered a death far worse than what we've seen. Now, do you know why He died?"

"Because of our sin. He died to save us from our sin."

"That's right. You and I are living in a fallen, sinful world, Smitty. It's a place where planes fall out of the sky, where people suffer and die horrible deaths. It's a place that has no guarantees. It might happen to us today driving home—or tomorrow. Who knows! But we're going to die, you and I. And it's better to face it now. The Lord Jesus Christ died on the cross to save us from our sin. On the third day—"

"He rose from the dead," Smitty interrupted.

"And why?"

"To give us eternal life."

"Right. Jesus said, 'I am the resurrection and the life; he who believes in Me shall live even if he dies' (John 11:25). Smitty, we don't have to fear death anymore, even if we die like those people on Flight 427. If we believe in Jesus Christ as our Savior and Lord, we will live—even when our bodies die. We will be with the Lord forever, having eternal life. That's His promise. That's why communion is so important. We eat the bread, drink the wine, and remember that Jesus Christ has faced death. He has forgiven our sin. He has been raised from the dead, and that means that believers have hope. Death isn't the end for us."

Smitty sat back in his chair, cupped his hands, and put them under his chin. He kept nodding his head, as if he understood. Then he said, "That's why He was there with me at the crash site, wasn't it? I didn't have any fear then. I knew He was there. I kept asking for strength, and He kept giving it to me every time. As a matter of fact, I've never felt closer to the Lord than I did that week. You're saying that's what's going to happen when I die, aren't you? He's going to be with me—like he was at the crash site. And He's going to give me strength to get though it. Jesus Christ is going to save me for eternal life."

"That's His promise. So why fear the nightmares? What we saw was awful. It will stay with us for the rest of our lives. But we don't need to fear death—not anymore. Not with Christ in our lives."

He looked at me as if I was the first person to tell him that. I wasn't. He had heard the words before, in church. He had tasted the bread and the wine, but it had never made sense. He had never faced death or the fear of dying before. It was easier to push it away. But now, he couldn't. He needed to hear words—words with the substance of knowing God, His love, His suffering, His promises, and His comfort. And he needed to hear them from somebody who had been there—like he was — and who understood what it meant to face death. That day, Smitty heard and believed in Jesus for his life.

The Need to Speak

Doctors have it much easier. They know their priorities. Their job is to medically diagnose and treat our illnesses. It's an added benefit to find a doctor who is polite and has a good bedside manner—one who speaks kindly and shows genuine concern. But it's not a requirement. In an emergency, we need the best doctor, not the nicest. We need someone who will make the hard decisions. If our illness requires surgery—cutting us open and causing us more pain—we agree to it. In the end, it's the road to recovery.

Imagine doctors whose priorities are reversed. They major in politeness. In fact, they're so polite that they won't recommend anything that might upset us. At the time, it might make us feel better. But what about the larger picture? Do we really want to hear the words, "You're going to be just fine" when if something's not done soon, that won't be the case? Which is better, a nice doctor who sweet talks us or the best doctor who cares first for our health and our future? Okay, we want both. But what if we had to choose?

Pastors have it harder. We tend to mix up our priorities. Our job is to meet people in their situations and bring God's Word. It's helpful when the pastor is a warm, polite person. But in times of crisis, people need someone who's going to tell the truth, even if at first it hurts them, offends them, or makes them angry. They need to hear God's Word, which alone provides true, lasting comfort.

It's so easy for a pastor to reverse priorities. For many years, I majored in politeness. I watered down God's Word, opting not to upset people. Standing in the pulpit at a funeral, I found it easier to be sentimental—to tell moving stories, quote great sayings, and end with, "You're going to be just fine." I found that politeness works; people respond—especially if I've spoken with kindness. They feel comforted; they like what I said and how I said it. So why speak God's Word? Why risk offending people or making them angry when politeness does the job? Which should I be, a nice Christian full of sweet talk or someone who will tell it like it is without compromise? Which is better for our spiritual health?

For years, I chose to be the former.

The plane crash marked a turning point for me. From that time on, I've devoted myself to God's Word first, kindness second. The tragedy forced me to stop compromising. But it wasn't the first time I knew my priorities were out of line. A few years before the crash, the Lord made it quite clear to me that my politeness helped no one. In fact, it was empty comfort which, in the end, was detrimental.

It happened on our summer vacation.

Erilynne and I spend a few weeks in Maine each summer. One late August afternoon around five o'clock, we were at the beach, reading. It's a small, sandy beach—unusual for Maine's rocky coastline—set peacefully in a cove and protected from the big ocean waves of the Atlantic. All the sunbathers had gone home for the night, leaving just the two of us. Our chairs were at the water's edge. The tide was up, the northwest wind was blowing strong in our faces, and the water was restless. Erilynne saw it first.

"Look at that!" she said, pointing ahead and just to our left. "What is it?"

A flock of sea gulls was hovering over the water, all bunched together, making loud cries as if a fishing boat was coming in with a great haul. But there was no boat. Just below them, the sea was frantic—churning and boiling like hot water on a stove. I stood up to get a better view. Neither of us knew what we were seeing—and it was coming toward us.

In a few minutes, it was so close—just a few hundred yards from the beach.

"Those are fish," I exclaimed. "They're jumping out of the water! What's going on?"

Down the sea gulls came, catching the fish in the air. Two men in a small fishing boat had pulled up to get their lines in the water. They kept waving their hands, signaling other boats in the area. One man screamed, "Blues are running!" The instant their lines hit the water, they had a fish.

"Those are bluefish," Erilynne said. "They're chasing

pogies." Bluefish are huge—twenty pounds and twenty inches or more. They had found a school of smaller fish and were feasting on them.

"Right, the pogies are on the surface. They're the ones making the water churn."

"They don't have a chance between the blues and the gulls," Erilynne observed.

With that, the boiling waters came at us, with the pogies popping right out of the ocean, swimming as fast as they could. In a matter of seconds, they were right in front of us, caught in the waves and thrown onto the beach, at our feet, right under our chair! Ten, twenty, fifty, then a hundred small fish were flopping helplessly on the sand and groping for life. They had sought safety in the shallow waters close to shore where the bluefish couldn't swim. But now they were in trouble again. Up they came, until the beach was packed with dying fish. The gulls were still overhead, and the blues twenty yards off the shore—circling, waiting, hungry, driven.

I got up. I couldn't watch these poor fish die on the beach with no water! One by one, I picked them up and threw them back. It was a bit frustrating. For every five I threw in, twenty more landed on the beach! So I doubled my effort, hoping to save as many pogies from death as I could. I thought I was doing a noble work until, at one point, I looked over at Erilynne, who was sitting in her chair amazed.

"Hey, come on! Am I the only one who cares for these pogies?" I laughed.

"What are you doing?"

"I'm saving these fish from death! If I leave them here, they'll die!"

"And what'll happen if you throw them back to the bluefish? That's what you're doing, isn't it? They're running from the blues! And you're sending them back. That doesn't make any sense!"

"But—" No, she was right. These pogies were condemned

to die either on the beach, by the bluefish, or by the sea gulls. They had nowhere else to go, and there was nothing I could do about it. I couldn't rescue them. I stood there helpless as their little bodies struggled for life all around me.

"I need somewhere else to throw them," I called back. "Got any ideas?" She shook her head and smiled. I hated the thought of doing nothing, but what else could I do? The waters kept churning in front of us. The sea gulls flew in, landed on the beach, and started picking up the fish in their beaks. I went back to my chair, resigned. We stayed there until the blues moved on down the coastline.

I learned two significant lessons on the beach that day.

The first: My politeness helped no one. I couldn't see the larger picture. I thought my kindness was saving pogies. It wasn't. I was sending them back to the world of hungry bluefish—an empty gesture of comfort and detrimental to their health!

The same was true in my pastoring. Opting for politeness over telling God's truth was doing the very same thing. I was sending the suffering back to the world of loss, grief, and death, with all their questions, concerns, doubts, and fears unanswered. That's not comfort. My priorities were wrong. My politeness in the pulpit was having as much effect as throwing pogies back to the blues.

The second lesson: I had to speak God's Word without compromise. There was nowhere to throw the pogies. They were condemned to die by the beach, the birds, and the blues. The forces of the world on those little fish are the same forces we face every day in a world dominated by sin, suffering, the lusts of the flesh, the power of the Devil, injustice, poverty, rape, murder, sickness, and terrorism. It's all around us! If it's not one thing, it's the other. Is there any comfort? I could do nothing for the pogies. But is there something I can do—something I can say—to the suffering? Is there a place of safety? of rescue? of comfort?

Yes, there is; we do have something to say. We have a message that must not be watered down with politeness. There are true words, God's saving words, given to us in the Bible. These are words of lasting comfort, though people may not always want to hear them. Some might be offended and turn away, but it was time for me to speak His Word without compromise. I knew that in my head, but how was I supposed to put it into practice?

I didn't know until the fall of 1994 when I was at the crash site. Then I knew it in my heart, for there I stood in the deepest pit of suffering I had ever known. I cried out for the Lord. I needed Him, not a dose of kindness. I needed to know that He was there, deeper still. And He was. I had a message.

---

[1]Corrie ten Boom, *The Hiding Place* (New York: Bantam Books, 1971), 192-193.

[2]Ibid., 211.

[3]Ibid., 217.

[4]Ibid., 195-196.

Chapter Twelve

No Compromise

"My dad's dying," Michael said over the phone. It was past ten o'clock at night. He had just come home from the hospital. His seventy-eight-year-old father had congestive heart failure. "Dad's lungs are full of fluid. He's in and out of consciousness. Both doctors met with my family this afternoon; they said that his body is worn out, and that it's best to keep him comfortable. Death is close; it could come today, maybe tomorrow.

"It's hard. Dad and I haven't had a good relationship," Michael continued. He'd been a friend for many years, but distance kept us from being close. I knew nothing about his relationship with his dad. "I'm the last of four boys. I came late. Dad was forty-two. By the time I grew up, he didn't have much time for me. He worked at his shop twelve hours a day, six days a week, and played golf on Sunday."

"Did it get any better when you became an adult?" I asked.

"Not really. Dad isn't a friendly man. He rarely smiles, doesn't talk much, never calls. The only time I remember him being affectionate was at our son's birth. He actually put his arm around me as I held Joshua for the first time—and chuckled! I'll never forget it. He was genuinely pleased to see his tenth grandchild. I think it was the first time I ever felt Dad's love—first and last. He isn't that kind of man."

"Michael, I'm so sorry."

"I'm scared. I'm really scared."

"Scared of being without him?"

"It's not that. I'm scared of what's going to happen to him. Dad only went to church when he had to. As far back as I can remember, Dad opposed religion. It was for sissies. Jesus Christ is the name he uses when he swears. Mom took us to church. I was baptized and confirmed as a Methodist. In college, I went to a Billy Graham crusade and received Jesus Christ as my Savior and Lord. He changed my life. That next summer, I told Dad what had happened to me. He didn't want to hear it. He called it nonsense. He told me I'd get to be his age one day, and then I'd know that it's all superstition—an old wives' tale. He didn't want to discuss it anymore. Two of my brothers tried to talk to him, but he wouldn't listen."

"And now he's dying," I said sadly.

"I don't know what to do. I love him so much. I don't want him to die without knowing the Lord. I've prayed for my father every day since college. When Dad was diagnosed with a bad heart two years ago, Carla and I decided we'd try to talk to him again. But it never worked out. The right time never came. And now, I think it's too late. The doctors said he might even die tonight."

Michael and I prayed over the phone. He wept for his dad, saying, "Father in heaven, please give me one more chance to talk to him—please. I don't want him to die without You. Save him, Lord. Open his heart like You did mine. Please!" His passionate voice was full of love and longing for his father. It was urgent. Death was coming. His father had lived life without God. And now, he faced eternal life without Him, and Michael knew it. He decided to go back to the hospital at seven the next morning.

Two days later, Michael called.

"I got to his room before seven. I couldn't believe it: Dad was awake, half sitting up in bed, and drinking juice. He was weak, but the night's sleep seemed to have revived him. As I sat at the edge of his bed, I knew the Lord had given me time. Mom wasn't there. There were no visitors or doctors or nurses. It was just the two of us. I went for it. I said, 'Dad, I have to talk to you. There's something bothering me.'

"He looked at me," Michael continued, "like he was concerned and wanted to listen. So I asked him, 'Dad, what's going to happen to you when you die?' And he said, 'I don't know, son. Nothing, I guess.' I said, 'So, you don't believe in heaven or hell? You don't think one day we're all going to stand before the Lord and give an account for our lives here on earth?' Although at one time it seemed like he had all the answers—confident and belligerent—now, knowing he was dying, he looked at me as if he wasn't sure he was right.

"I said to him, 'Dad, you'd listen to me if I told you about a good car to buy or a stock you could make some money on. Well, listen to me now.' So I started at the beginning with Mom taking us to church. I told him what happened to me at the Billy Graham rally and how Jesus Christ saved me, a sinner. I said right out, 'Dad, He came to save sinners, and He can save you, too.' You won't believe this, but he started asking all kinds of questions: How did I know? Why did Jesus die? Is the resurrection really true? Can anyone be sure he's really going to heaven? He said, 'How could God ever forgive me for all I've done in my life?'

"Thad, we talked until half past eight. At the end, he asked me to pray for him that Jesus Christ might forgive his sin and save his life too. My father! On his dying bed! The Lord saved my dad!"

Michael's father lived another week. He was different, Michael said. He kissed his sons, held the hands of his grandchildren, and told his wife for the first time in decades that he loved her. "It was still Dad," Michael said without much emotion or fanfare, but there was a change. He was loving and being loved as never before. Michael's grief after his father died was mixed with thankfulness. He knew that his prayers were answered. His dad was safe with the Lord in heaven, and that made such a difference.

I've never forgotten Michael's story. That first night he called, I had heard something I rarely hear; the man was driven to pray for someone who was lost, without God in this world. He was scared for his father's soul. The situation was critical, urgent. Time was pressing in. He needed to pray for time to speak with his dad—before it was too late. Michael had a passion for the lost and the drive to do something about it.

Burden for the Lost

I saw that same passion in the Third World.

Seven months before the plane crash, Erilynne and I were in Uganda, East Africa, on a preaching mission. We were driven a hundred and twenty miles northwest of the capital city of Kampala to a town called Bulindi, far away from First World comforts. Electricity was sparse and irregular. People walked miles to get their five-gallon jugs of water filled, and even then it was undrinkable water; it had to be boiled first. There were few cars on the bumpy dirt roads—mostly bikes and the occasional white taxi vans that were designed to seat eight but packed in eighteen. The people had access to few paying jobs, few stores, and few doctors and nurses with little resources to give decent medical care.

We met with many of the local Christian clergy and lay leaders, and realized the sharp contrast between us. These men and women were on a mission, as if a military operation were in place. They were serious, devoted, clear about their message, and ready for action. We felt their urgency but didn't understand it. We had the same work and the same message, but not the same passion. Where did it come from?

"We are a people who know death," an Anglican priest in the Church of Uganda told us. "During the seventeen-year military reign of Idi Amin and Milton Obote, more than one and a half million Ugandans died. Since 1986, our new government has brought peace to our country, but death continues to reign—death from poverty, malnutrition, sickness, disease, hostility, childbirth, orphaned children who are left to die, alcohol, and now the deadliest killer, AIDS. Soon AIDS will kill more than those who died under Amin and Obote. Death is constant for us—all ages, anytime, any place. Our condition is serious; today is all that many people have, for tomorrow they may die. So we must act right away."

"What do you mean?"

"Our people must not die without Jesus. Only His blood that was shed on Calvary can save their souls at the day of judgment. We must tell people what He has done before they die, before the door of heaven is closed to them forever. They must not perish in hell. Jesus Christ called it 'the outer darkness; in that place there shall be weeping and gnashing of teeth' (Matt. 8:12). One soul whose name is not written in Jesus' book of life is our burden. We don't want them to end up in 'the lake of fire,' as the Bible calls it (Rev. 20:15). So we must tell our people about Jesus Christ and what He has done to save us from sin and death. He can save them, too."

This priest had a burden for the lost, and now I knew why. Their situation in Uganda was critical. Death was always near. Time was short. Something had to be done—*now*. Being a

Christian meant that these men and women had met Jesus Christ as their Savior. He had forgiven their sins and rescued their souls from hell, for heaven. Now, filled with love for their own people, these Christians, young and old, poor yet possessing the riches of God's kingdom, were on a mission to tell others—before it was too late.

It was just like Michael going to his dad.

I didn't have that passion. I live in the First World, where the situation is not critical. Death isn't pressing in from all sides. In America, we fight crime, drugs, and injustice, but we're not at war. Many suffer from poverty, but the majority live in plenty. Sickness, disease, cancer, and AIDS are with us, but we have no lack of doctors, medication, hospitals, or the most sophisticated technology in the world. Most of us have a roof over our heads, meals on the table, good water, cars, decent paying jobs, health insurance, a retirement package, money in hand, and a sense of security that tomorrow we'll be here! We assume we'll be safe, healthy, and living without fear. We think we have time; in fact, we have so much time that we rarely think about death.

Where's the urgency? What about the drive to tell others about Jesus Christ? I believe we don't have it because we're not in crisis. Death isn't looming oppressively over our heads. Why speak with passion about sin, judgment, death, and hell—eternal punishment, weeping and gnashing of teeth, a lake of fire—or of a Savior who offers a real and glorious heaven? It's a message for those who are facing death and know their eternal destiny is in jeopardy. On the whole, that isn't our American culture. We don't live in a time of crisis, so we don't talk about it—not seriously—not until we're faced with tragedy. The situation has to be critical; death has to be near and the time short. Then, in our hour of need, we consider God. We ask our questions.

Part of me didn't want to leave Uganda. Living near death, around so many people who had one foot in the grave, gave me a new heart for the lost. I saw their crisis. I felt the urgency to act by telling people about Christ and His cross. That fire was still in my belly when I came back to the United States. But after a few months, it slowly fizzled out. I was back to my normal routines. There was no urgency, no "before it's too late" mentality.

Then the plane crashed. Standing at the impact site brought it all back. Was there one lost soul on board? What was I thinking? Where was the passion and the urgency? Why had I allowed it to drain from me? Don't Americans die, too? Isn't death right here, and isn't our condition serious? Isn't the message for Michael's dad and for Ugandans the same message that's needed in the U.S.? The rubble and the death all around me was proof enough. But I had forgotten; I had lost my drive for winning the lost.

And it scared me. These people didn't have time—no time at all.

Their situation was never critical, not until the last twenty-three seconds. One moment, death wasn't near, and then without warning, it was suddenly upon them. There was no time to think about dying, no time to talk about their eternal destiny, and no time to hear about a Savior who had come for them. It was over, just like that. Were there people on board that plane who were like Michael's father—people who had lived their lives without God, bound in sin, wrapped up in themselves, and destined by their own choices for the lake of fire and not the glories of heaven? Maybe not. Maybe so. The possibility disturbed me deeply.

"One soul," said the Ugandan priest, "whose name is not written in Jesus' book of life is our burden."

The struggle flared inside me. Why had I lost the passion and urgency to tell people about Jesus Christ before it was too late? Had I fallen prey to my own culture? Was I blinded from seeing death because I live in a First World nation? As a Christian, I had lost the drive. As a preacher, I had lost the clarity of the message. At the impact site, I couldn't help but blame myself. I didn't know anyone on board that plane, but I knew people like Michael's dad who weren't ready to die. So why hadn't I talked to them about the Lord? Was I scared—afraid they'd reject me? But what if they had been on Flight 427?

I needed the passion again—not just then, but always. I needed the urgency—not for a moment, but forever. I begged for it: "Lord Jesus, let me live for the lost. Let me live to tell them about You."

No Struggle, No Real Comfort

A retired Lutheran pastor in North Carolina saw my picture in the newspaper a few days after the crash. He chased down my address and wrote me a letter. I got it after my work at the site was over.

"I know what it's like to be there," he wrote. "I was in Vietnam as a chaplain from '66 to '68. I saw a chopper loaded with twenty-six soldiers explode twenty feet off the ground. We worked at that crash site for five unforgettable days. When the USAir plane crashed outside of Pittsburgh, it brought back memories. I saw your picture in the paper. I thought you'd like to hear from someone who has been in your shoes."

He was evidently a loving and caring man, and his letter was just the right medicine. He wrote:

> We never found out what happened. It looked like a bomb to me. But our military experts said the explosion could have been a mechanical failure. I never heard the final report. I thought it was odd I was watching the helicopter. Why wasn't I in my office or out in the field with the soldiers? Instead, I was standing on the tarmac, concerned for one of the men on the aircraft. For a week, he had been having nightmares about dying in combat—a grenade explosion. He didn't want to die,

afraid he'd never see his wife or two-year-old daughter again. So he came to my office. We met every day that week for counseling and prayer. When he got his orders to go to the front line, I felt I had to be there to see him off. Little did I know, I had gone to see him die. That explosion, those flames, killed my friend.

The bodies were beyond recognition. Most of them were not intact. All of them were burned. The crash site had a radius of nine hundred yards. The blast ripped the chopper to pieces so it took a lot longer to clean up. Our job was to get the bodies to the morgue and pray the men had left something behind for their families—a ring, a photo, anything. I found the body of the young man I'd known. He carried his diary in his top left shirt pocket. Funny, it survived. His last entry was a note to his wife which ended in a prayer. Not for his safety. But for the Lord to keep and bless his wife and daughter.

The military awarded me with a medal of bravery for working at the crash site. What a twist. Bravery described my friend—not me. He spent his last week knowing he might die in battle. And still, he boarded the chopper. That, to me, is bravery. I sent the diary to his wife with a long letter telling her the details of his last days, and how much he loved her. A few years after the war, I flew to Chicago—where she lived. It was my distinct pleasure to give her the medal of bravery her husband deserved. All these years later, and I still remember his face, the fear in his eyes, and the last prayer in his diary. In a small way, I helped him prepare for death. In a big way, his courage has given me strength to live.

After all these years, this Lutheran pastor wrote as if the explosion had just happened. The images were that sharp, the reality of death that clear. I knew the same would be true of me om now. To see death—I mean really see it—is to live li[illegible]erently. For both of us, it had changed us forever.

As I read on, I secretly wanted him to feel the same way I did. Did he have the same questions? Was he struggling as I was? How did he deal with the deaths of those twenty-six men? Did he wonder whether they were ready to die and meet the Lord? Was it possible that even one of those souls died without knowing God? Did that bother him? Did it give him a longing to tell people about Christ? And now, after all these years, was that longing still there? At the end of his letter, he dealt with my questions:

> So what advice can I give you? I imagine you're living with the sights of the crash scene every day, going over and over them in your mind. You're probably thinking about those who died, and those they left behind. Caring for their remains made me grieve like they were part of my family—was it the same for you? Before I left Nam, I was waking up at night with sweats. It was hard to counsel soldiers who, afraid to die, needed to talk. It was even harder conducting funeral services. The explosion, the fire, the five days working with the dead bodies, were always on my mind. I didn't know what to do.
>
> After I got home, I sought help from a military chaplain. He gave me some advice I've never forgotten. It's worked for me. I hope it works for you. He told me to pray and picture those twenty-six men in my mind. See them, not strapped in the helicopter moments before they died. But see them alive, happy, smiling and standing in heaven with the Lord. Don't remember them in agony. Remember them as they are right now, safe in heaven. So I did, every night in my prayers before I went to sleep. It took a while, but the night sweats disappeared and I thought less about the

> explosion and my days at the crash scene.
>
> I recommend this kind of prayer for you. It will ease your grief and help you to live out the rest of your life in peace. I don't know how people face death without the Lord's promise of heaven. There's too much suffering and anguish in this world. I'll leave it at that. Know that I am praying for you.

He signed the letter, "A fellow sufferer." I was moved by his letter and his compassion for me.

The last bit of advice, however, didn't help. It wasn't fair. In his effort to seek comfort, he had opted for an easy path. He saw all twenty-six men "safe in heaven." That eased his grief. But was it true? Were all twenty-six men with the Lord? How did he know that? Was it a guess? Had he known them personally? Did they attend his services? Did he pray with them? Or was it blind, empty hope? Was it what he needed to comfort his own pained soul? Clearly, the thought itself calmed his night sweats and soothed painful memories.

But was it false comfort? What if he was wrong? What if there were men on that chopper who had rejected the Lord and died without Him that day—one man, five men, fifteen—what then? What if we pictured some of those men, not safe in heaven, but unsafe in hell? What does that do for us?

For me, this was the heart of my struggle. To think that one person aboard Flight 427 died without the Lord drove me to my knees. I needed those night sweats and the painful memories of what I saw at the crash site. They taught me that life is short. Death is certain. Our eternal destiny must be addressed now, not later. Therefore, I must be concerned for the lost today, so that they might not die without Jesus Christ. I refused to be comforted in my grief. It was the wake-up call I needed to take seriously the mission of reaching the lost before it's too late. That grief in my heart revived the passion and the urgency I had once known.

I couldn't pray the pastor's prayer. He might have been right that all of those aboard Flight 427 may have known the Lord and are "safe in heaven." I don't know. But this I know: Jesus Christ came to save sinners. He died on the cross and was raised from the dead so that all who believe in Him might not perish in hell, but have eternal life. That's the gospel. People without Christ need to hear it now because death can come at any time—even in twenty-three seconds—whether they're ready or not.

No, I couldn't take the easy path and make believe that all those who died are in heaven. It might have brought comfort, but is it true? And if it isn't true, do we opt to keep the false comfort or do we face the hard facts of life—that there is a real heaven, a real hell, and a real need to tell people while there's still time?

The Memorial Services

I read and reread the pastor's letter. There were days when I envied the comfort he'd found. It seemed to bring closure to that dark, tragic time. The memories were clear, but they didn't have the same power or wield the same control. I wanted that—but not his way. Still, I understood what he had done. I didn't agree with it, but I knew firsthand the suffering and the anguish of his heart. I knew he needed relief.

For him, it worked. But again, it's like a doctor giving morphine. It "works" because it relieves the pain. But it doesn't cure, and to think it does is wrong thinking. To offer it to others as a source of comfort is just as wrong. With grief, morphine comes as a quick fix, a simple "do this and all will be well." But all isn't well. This kind of comfort masks the depths of a hurting heart. I didn't want it. I couldn't receive it, nor could I give it to others. But the pastor's love, that I received with gratitude. He genuinely wanted me to feel better.

His letter came a few days before the September 25

memorial service. This would be the second public event of its kind. The first one was held in downtown Pittsburgh the Monday following the crash when the rescue workers at the crash site and morgue couldn't attend. So the Salvation Army planned a large, outdoor service to be held at the Green Garden Plaza the following Sunday afternoon at half past two. Their first concern was to bring comfort to the family and friends of those who had died on Flight 427. But they also wanted to reach a host of others—the county coroner and his staff, the hundreds who had participated in the rescue operation, those who lived in Hopewell Township, the men and women of USAir, those who worked at the Pittsburgh airport, and anyone else mourning the tragedy. It was to be a time just to be together, to grieve, and to find solace.

The Plaza was just the right spot. The rolling hills were in view, as was the vivid memory of that first night when black smoke filled the sky, sirens roared, and thousands watched and hoped for survivors.

Three thousand people attended the service. The governor of Pennsylvania, Bob Casey, flew in to speak to the crowd. Church leaders from all denominations were invited to participate. Choirs sang, bagpipes played "Amazing Grace," Bible passages were read, prayers were offered, and the names of the 132 who died in the crash were spoken, one by one. The song "As We Sail to Heaven's Shore" was played over the loudspeakers. Kirk Lynn had recorded this song on August 28 at his home church, eleven days before the plane went down. Now, weeks later, his voice—like the voice of an angel—deepened our sense of loss. He wasn't there to sing it in person—the song that called us to think of heaven. Kirk Lynn, age twenty-six, was one of the 132.

Three of us were asked to give brief sermons. As I prepared to speak, I remembered a television interview with a retired newspaper reporter. He had clear advice for young reporters: "Tell the facts! Stop telling people *what they want to hear.* Start telling them *what they ought to know!*" I saw that same objective as my first and only priority. I was given five short minutes to speak. I knew I had to keep my focus set: *Remember those deeply grieving.* Tell the facts: *Jesus knows the pit of suffering firsthand. He alone has the power to comfort the soul in the shadow of death.* That's it. Anything less would be a quick fix of morphine.

I would give just the facts—no compromise. I had so little time.

One subject was forbidden: eternal destiny. When Christians die, people weep and mourn for their loss, but there's something different about them and their grief. They know that their loved ones are with the Lord Jesus Christ in heaven. That comforts the soul in sorrow like nothing else. And more, their grief is filled with a certain hope that one day they'll be in heaven, too, with Christ and their loved ones forever. They know this even as tears stream down their faces. For Christians, the subject of eternal destiny is a fresh, soothing spring of hope.

When nonbelievers die, that hope isn't there—nor can I, the pastor, mention it. I buried one man who went to his grave cursing God despite his daughter's faithful witness and constant prayers. The family asked me to speak at his funeral. His daughter made this request, "Please tell our family about the Lord." And I did. At a nonbeliever's funeral, my job is to comfort the living. For them, it's not too late. They can turn to Jesus Christ and find help from Him. Why speak about the eternal destiny of the nonbeliever? It can only lead the family to despair—and it might even block them from finding Christ in their loss.

Of course, I must always tell the truth. What if I said to the family, "Don't worry, he's in heaven with the Lord?" I'd be guilty of telling people *what they wanted to hear,* even though it contradicts the Bible. Will I say anything for the sake of comfort? I don't think so. At a nonbeliever's service, it's best to leave the subject of eternal destiny alone and focus on the living.

That's what I did at the Green Garden Plaza memorial service. Were all 132 people on Flight 427 believers? I didn't know. None of the ministers knew. None of us could honestly promise, "Take comfort! Your loved ones are in heaven with the Lord today!" But it sounds so good, and it makes people feel good. *Isn't that the point of a memorial service?* some would ask. *So why not say it? Why not give people a ray of hope?*

And to my surprise, it happened over and over again at the various services. Ministers from a wide variety of Christian denominations, some holding high and respected offices, stepped up to the pulpit and promised, "You must not grieve without hope. God is a loving God. He has not forgotten your loved one. In their fiery trial, God heard their cries and took each of them home to heaven and into His everlasting arms of love."

"I don't understand," I said to one of the ministers later. "How can you make that promise?"

"Because I believe it. I believe that all who die go to heaven."

"But that's not what the Bible teaches. Jesus said there are two gates. One is broad; it leads to destruction, and many go that way. The other is small; it leads to eternal life, and few find it (Matt. 7:13-14). He also said that there's a real hell where 'the fire is not quenched' (Mark 9:44,46,48). The Bible never says that all who die go to heaven."

"Look, I believe God loves us all, and no matter what we say, what we believe, or what we do, God will forgive us in the end. I think everybody gets saved—regardless of what the Bible says."

"If you don't believe the Bible, how do you know you're right? What do you base it on?"

He pointed to his heart and smiled softly. With that, he shook my hand and went his way. I still didn't understand. How can he be a Christian, let alone a pastor, and not believe the Bible? And how can he believe that all are saved? If that's true, then why don't we freely sin? Why not murder, hate, commit adultery, steal, lie, and live as we please? If there's no punishment for sin, what holds us back? Let's live it up! Why follow the Lord? Why choose right over wrong and good over evil? What prevents total anarchy?

If all go to heaven, then I don't need a Savior. So why did Jesus come? Why did He bear our sin in His body, suffer on the cross, and die for us? Why did He command us to repent of our sin, believe on Him for salvation, and live godly lives on this earth? Why did He give His church the mission to preach the gospel to the lost? If all are saved, then there are no lost people. There is no gospel to preach. What happened on the cross means nothing. We're left to say what we want, believe as we choose, and do as we please.

I felt as if the wind had been knocked out of me. I could feel the temptation inside: *Why struggle so much? Why not give in? Tell the grieving, "It's going to be okay. You'll see your loved one again in heaven. We're all going there someday!"* It's what they wanted to hear. It would bring them comfort. It would make them feel good—and isn't that what's most important? So why not take the easy path?

Why not tell them what they wanted to hear?

Five Minutes

When I stepped up to the pulpit at the Plaza, I felt like Michael sitting at the edge of his father's death bed. I felt like the Ugandan priest who had one passion: to tell people about Jesus before it's too late. I couldn't take the easy path. In that crowd, there were people grieving. For them, the situation was critical. They needed answers and comfort—not morphine comfort, but the real stuff, the real Savior.

It was time to tell the facts. I began with a brief prayer and started to speak:

I have come this afternoon with these few words to the

family and friends of Flight 427 and to those who worked so hard at the crash site, the morgue, and in the rescue operation. I have one thing to say: There is comfort to be found here today in this hour of grief. Real comfort!

The Comfort of So Many People

Through the kindness of the Salvation Army and the county coroner, my assistant and I were allowed to serve at the crash site. I can tell you, on my first ride up that Friday morning—some fifteen hours after the crash—I thought to myself, *What am I doing here? It would be far better to stay at the Plaza and tell the rescue workers going to the site, "You go to the top! Know we're praying for you, we'll be here when you come down."* See, it was hard to be there—for all of us. It's easier not to face tragedy and stand at a distance. But that's wrong. We had to face it—together.

I heard people say up there, "It looks just like war." But I must say this: We can never forget, war is what we do to each other. It's the human race at its most vile. We divide in brutal hatred against each other. This was not war. This was an accident. The hardworking people of Beaver County, the great city of Pittsburgh, the surrounding counties in western Pennsylvania, and the United States government personnel were not divided—but united, working side by side as one person. Volunteers joined the rescue operation from every walk of life. They came to help, to give their time and service, to grieve, cry, and stand with you, the families, in your hour of need.

Families and friends of Flight 427, look around you. See the faces of hundreds of workers from this county, this city, and this nation who were here from Thursday night on—ready to serve, ready to fight for even one survivor, ready to care for you and for your dead. It wasn't possible for you to be at the crash site. But these people were there—honoring you, honoring the dead, and honoring the Lord by their work. Know their love for you today, and take comfort from it.

The Comfort of the Lord

But there is a greater, lasting comfort.

Working at the crash site, I knew the promise of the Bible, "Yea, though I walk through the valley of the shadow of death, I will fear no evil. For Thou art with me. Thy rod and Thy staff they comfort me." I wish the Lord had promised that when we're in the valley of the shadow of death, He'd take us out. But that isn't it. His promise is to be there with us, to stand with us so we might not fear evil. His promise is to stretch forth His rod and staff that we might know His comfort.

That's what I did at the crash site—I told rescue workers the Lord our God does not stand outside the tragedies of this life. His promise is to stay with us. The greatest comfort that can be offered in this broken world today is this: Our Father in heaven doesn't back away from our grief, our trials, our sin, our suffering, our death. *He doesn't back out!* He never has. He never will.

If He did, He Himself would never have come that Christmas night to the little Judean town of Bethlehem. God became one of us. He stepped into our experience as fully man. He never backed out! And more, He would never have gone to the cross of Calvary to take to Himself the sin of the world. He'd never have suffered our pain, our sorrow, our death for us. I'm here to say, Jesus Christ knows your suffering firsthand. He will stand with you. He will bring you comfort.

So take heart! Take courage, all who mourn! For the Lord Jesus Christ whom I serve will be with you as you call on Him. He will stand with you in this dark hour of grief. He will be there in the valley of the shadow of death. He who sits at the right hand of His Father in heaven today is ready to meet you and comfort you as you call on His name. Turn to Him! Believe in Him! Know He will be your help in this life. And when you face death, you'll know the triumph of Easter day and the promise He alone can give of heaven: *Jesus Christ has conquered sin and death. He is risen from the dead!*

To the rescue workers who need comfort today, who remember the scenes of the crash site or the morgue in your thoughts and dreams: Call upon the Lord. I know what it's like. I was there. I know how hard it is. And I know Jesus Christ alone has the power to comfort our soul.

To the family and friends of Flight 427, receive the comfort of all of us surrounding you here today. And receive the comfort—the real comfort—that comes from the Lord. He will not back away. He will meet you in these days of suffering. His rod! His staff! That's His promise. So turn to Him and put your trust in the Lord and Savior Jesus Christ this afternoon. Amen.

Chapter Thirteen

FACING LIFE

One year later on September 8, 1995—the anniversary of the crash—three services were held. The first took place at the soccer field, which a year before was crowded with children, coaches, and parents getting ready for the fall season. Some looked up as the 737 went overhead, not knowing it was the last time the plane would be sighted. The second was held at Sewickley Cemetery, where many of the 132 are buried. The final service was held up at the crash site, beginning at seven o'clock sharp. It was hoped that we'd all be standing at the impact site in silence at 7:03—the exact time Flight 427 hit the ground.

Buses started leaving the Plaza at 6:20. Word had gotten around that this service would be the final public event to be held at the crash site—one last opportunity. No one expected the crowds. Bus loads of people were still coming at a quarter past seven. I had been asked to lead the service. We started at seven o'clock sharp, but there were too many people. Some missed it altogether. They asked that we repeat the service, and we did as dusk turned to dark.

Being there again opened a floodgate of memories and emotions for us. The land had healed faster than we had; there were signs of new growth all around. Some trees near the impact site were still scarred, with dead branches up to the top. But on close inspection, even they had traces of new life that revealed an urge to live. On the bark and scattered on the ground, I saw a few wooden plaques and crosses, small and unassuming, each bearing the name of a victim. A passerby strolling down the logging road would have had to look carefully to see them—or even to notice that once not long ago, tragedy had struck here—a jetliner had burst into flames, people had died, and the world had looked on with concern and sadness.

I saw a sea of faces that night that I hadn't seen in a year. Some faces were seared into my memory. In a flash, I'd picture them in my mind or in late night dreams, and I'd feel the aching grief again. Over time, I'd wonder if those faces were real or made up. Is that how he looked? Did her voice sound like that? What was his name? The line between the real and the imaginary had become fuzzy, but there they were! There they all were at the impact site, as if time hadn't passed and the plane had just gone down. It wasn't a dream at all.

Erilynne touched my arm and pointed to her watch. It was seven, time to start. I asked the people to stand where they were and opened with a prayer for the Lord's blessing on our last time together at the site. We spent the next five minutes in silence. I kept an eye on my watch: 7:02, 7:03 . . . .

At first, the only sounds were of people weeping. But then, at the precise moment of 7:03, chills went down my back. That sound—overhead! I looked up; we all did. Directly above us was fifty feet of open sky, not blocked by trees behind us or to our right and left. There, in that little window of sky, was a dark shape behind the white clouds. I heard the sound of engines and a gasp in the crowd. It was a plane! A plane was taking the same flight path at the same time; it was even the same kind of plane, with one engine on each wing, just like the 737. It flew by, and as it did, more people began to cry. We stayed silent until the sound of the engines faded slowly from our hearing.

It wasn't planned; I later asked Terry Thornton, assistant manager for operations at the control tower in Pittsburgh, about it. Some thought it was eerie. Others called it a coincidence. For me, it was a cold, hard fact. Time marches on, no matter what we suffer or how defiantly we protest. People resume their lives, planes still fly, and routines continue regardless; it's business as usual. Nothing stops the persistent passing of time. It's just so hard to keep pace when one moment in time, 7:03 to be exact, hurt so much. There we stood, one year later to the second, and from above came the blatant, callous reminder that time never freezes, never waits, never cares.

I've had to face facts, I told the crowd, that time is short. I learned that from the crash. No one holds the guarantee of tomorrow. There are no sure promises—not for the youngest, not for the healthiest, not for the optimist or the wealthy, or for those who do good deeds, or even for the faithful believer. Each one of us knew what "twenty-three seconds" meant: One minute you're alive with plenty of time; the next minute you're gone. Time is short because this life is fragile and full of crisis, sickness, sin, evil, accidents, and trauma—the constant, unpredictable storms of life. As for me, I told them, I used to live assuming that I had a tomorrow. No more. The "twenty-three seconds" changed all that.

Facing life is different. With time on our side, I said, we've got certain priorities. For some, that means success: a big career, making lots of money, having the right house full of nice things, a good car, an "in" wardrobe, or enough attention. For others, the priority is pleasure: weekends away, hobbies, sports, hours of TV—anything to fill up the time with leisure and fun. For both types of people, it's too easy to sacrifice the most treasured of relationships for the busy life that success and pleasure demands. Isn't it odd that when we know our time is short, it all changes? Priorities shift; what we once valued turns cheap, and what we once took for granted becomes everything. When we know we have no time to spare, we can't get enough.

As for me, I told them, my relationships came back to the center. It's amazing, isn't it, that suddenly nothing compares to holding our children or being together with our spouse. There's a longing to rekindle our love for our friends, even those friendships we've left behind for no good reason. They all matter, now. Before, they may have mattered in principle, but not in practice—not in actual time spent. And no relationship has a chance without time spent. Too often, we learn that lesson too late. It's better to put it into practice while there's time. I learned that from the crash too, and it has made a difference.

One more thing, I told them. Facing God isn't the same either.

The Promise of Trouble

Henry was on a bus in East Africa, traveling with friends. The bus was ambushed by guerrillas firing guns. Henry was shot, and half his face from below the nose was blown away. It was nothing short of a miracle that he survived. World Vision, a Christian relief organization, paid all the expenses for a hospital in Montreal to rebuild his face. David Watson met Henry there. "I could not help flinching," he later wrote, "when I saw the mask that had once been a face." But the African's eyes sparkled, and his heart belonged to Jesus. He couldn't speak, so Henry wrote David these words: "God never promises us an easy time. Just a safe arrival."[1]

That's not what we want to hear, nor do we tend to believe it. Christians go to church, read their Bibles, pray with their families, and expect God to answer their prayers and bless their earthly lives. We expect health, protection, and safety—especially for our children—daily provision for our physical needs, and security in knowing we've got tomorrow. We finish our prayers, confident that no harm can come to us and our family. The Lord is on our side. And He, we believe down deep, promises us an easy time.

The same is true for most nonbelievers. When crisis strikes and they can't handle it, they offer a prayer to God: "Oh, God, help me!" The situation demands that they consider the Lord. If only He'd relieve them of the crisis and make life easy again. It's understood, by believers and nonbelievers alike, that the Lord's job is to give us an easy earthly life. It's no wonder that when facing tragedy, both have the same question: *"Lord, where were You? Why did this happen?"* It gives "proof" to nonbelievers that God isn't loving, good or fair to allow this tragedy. So why worship Him? And it frazzles the minds of believers, who feel they deserve protection.

I heard a minister illustrate this point in a sermon. His wife had given birth to their first son. That night in the hospital, they thanked the Lord for blessing them with a healthy child. In the bed next to them, a woman answered the phone. She had delivered two months prematurely, and the baby's life was still in danger. Her husband was out of town. She hadn't even seen her daughter yet, let alone held her. She was afraid for her child's life, upset she'd have to leave the hospital without her, missing her husband, and physically drained. Her voice picked up when she heard his voice over the phone.

"See, honey," the minister said quietly to his wife, "I'm so glad we're Christians! Listen to that poor woman. She's having so much trouble. If only she and her husband believed in the Lord."

"I'm all right," the woman in the next bed said to her

husband. "I miss you. I want you here. I want our family to be together. But I know it's going to be okay no matter what happens. I've been spending a lot of time in prayer and reading my Bible. I've had many calls from our church friends. I know the Lord Jesus is here with me, taking care of me and our baby. He will see us through. I know He will."

The minister's face blushed red with embarrassment. How could a Christian experience such real suffering? He had thought, as so many do, that God promises an easy life for those who believe in Him. Even as he gave his sermon, I could tell he was visibly moved. This young mother had shown him two things: God doesn't promise "easy," but He does promise He will be there no matter what happens.

Many American preachers have fought hard to defend this "easy" view. It's popular; people want to hear it, and it's a positive approach to life. They say, "If you're sick, have faith; God will heal you. If you're facing a crisis, suffering from a broken relationship, financially poor, or in any need at all, just pray with faith to the Lord and He will rescue you. With prayer there can be no more tears, pain, suffering, sickness, poverty, crisis, or personal need. God wants us to enjoy this earthly life. We need to believe that and receive it to ourselves by faith!" No wonder television preachers have become famous with this type of "faith" message. It sells. People want to believe in a God who promises no unfortunate surprises and no suffering.

Imagine sharing this good news to the young mother whose baby is fighting for life in a hospital incubator. Or tell it to Henry, who was shot by guerrillas—or to the family members of Flight 427 passengers. It would never sell. People in the Third World who are afflicted with poverty, famine, disease, and war would never understand it—nor would anyone familiar with suffering in our part of the world. These preachers twist both the Bible and the minds of their listeners. And at one time, I believed them. I used to think that God promised "easy" until I faced suffering myself.

It's hard to face God in tragedy. We don't understand Him or what happened to His promise. Why didn't He protect us from the trauma? We storm and stomp with our fists raised in the air. We feel confused, betrayed, and angry. Was it us? Did we do something wrong? What if we had had more faith? Nothing makes sense; we only know how much it hurts. Our view of God hadn't prepared us for tragedy.

Henry knew the secret: "God never promises us an easy time."

The Bible never dodges the reality of our condition in this world. Jesus said that at the end of time, there will be wars, famines, and earthquakes (Matt. 24:7). Paul described the last days in this way: "For men will be lovers of self, lovers of money, boastful, arrogant, revilers, disobedient to parents, ungrateful, unholy, unloving, irreconcilable, malicious gossips, without self-control, brutal, haters of good, treacherous, reckless, conceited, lovers of pleasure rather than lovers of God" (2 Tim. 3:2-4). Living in this world isn't easy. It never has been. Until the last day, it never will be. Sin and suffering will never leave, and that's a fact.

On the night before He suffered, Jesus said, "In this world you will have trouble. But take heart! I have overcome the world" (John 16:33, NIV). He said it clearly: There's "trouble" in this world. And the very next day, He faced that trouble when He went to Calvary, bore our sin, and suffered God's wrath against sin for all time. To believe in Christ the risen Savior is to know we are forgiven. We're promised eternal life with Him who triumphed over death and overcame the world. In Him, we can "take heart"! It doesn't matter what happens in this troubled life; our hope is in Jesus Christ and His promise to be with Him forever.

Suffering and trouble are guaranteed, plain and simple. There's no escaping them.

Jesus Christ said to those who wanted to follow Him, "Whoever does not carry his own cross and come after Me cannot be My disciple" (Luke 14:27). To carry a cross is to know suffering as our Lord did, and that's the call of a Christian. The apostle Paul said in his letter to the Philippians, "For to you it has been granted for Christ's sake, not only to believe in Him, but also to suffer for His sake" (Phil. 1:29). It is wrong at the very center to believe that God has promised an easy passage on this earth. He has not!

Dietrich Bonhoeffer, a German Lutheran pastor, died at the age of thirty-nine. He was hanged by the Gestapo on April 9, 1945, just days before World War II ended. While in prison at a Nazi concentration camp, he wrote a poem titled "Who Am I?" His struggle in unbearable conditions is captured in this excerpt:

> Am I really that which other men tell of?
> Or am I only what I myself know of myself?
> Restless and longing and sick, like a bird in a cage,
> struggling for breath, as though hands were compressing my throat,
> yearning for colours, for flowers, for the voices of birds,
> thirsting for words of kindness, for neighbourliness,
> tossing in expectation of great events,
> powerlessly trembling for friends at an infinite distance,
> weary and empty at praying, at thinking, at making,
> faint, and ready to say farewell to it all.
> Who am I? They mock me, these lonely questions of mine.
> Whoever I am, Thou knowest, O God, I am thine![2]

Here is Bonhoeffer facing life. He knew this world and its suffering. His faith wasn't suffocated by the empty view that God promises "easy." How ridiculous! How shallow! His Lord had suffered, and Bonhoeffer knew that nothing less would be required of him. His faith was ready and present in the midst of "restless and longing and sick," "yearning," "thirsting," "tossing." It rose up and exclaimed, "I am thine!"

Christians, especially in the First World, must come to terms with suffering. All of us, believers and nonbelievers alike, have the same promise of trouble in this fallen, unstable world. We're fools to think that less will come to us. Now is the time—while there is time—to consider the Lord rightly. For nonbelievers, why wait for crisis and death to come near? What if there is no warning—what then? Why not turn to the Lord now? Seek Him, and as the Bible promises, find Him (see Is. 55:6 and Matt. 7:7-11). For believers, it's basic: Throw out the false view of God and of the world. Embrace the truth: There's suffering in this life, and we're far from exempt. In fact, Christians should expect more suffering as we take up our cross and follow Jesus.

"When Christ calls a man," Bonhoeffer wrote, "He bids him come and die."[3]

But this, for Christians, often means insecurity. Why pray if there are no promises? Why ask the Lord to protect our children when they leave home in the morning—or to heal the sick and help the elderly, provide food when we're jobless, grant mercy when we travel, and defend us in times of trouble? Why ask for grace before an operation or wisdom to resolve a conflict? What's the point if there are no guarantees?

But is that why we pray—to get the guarantee? Or do we pray to entrust ourselves, our loved ones, and our situations to the Lord and His care? Do we pray to avoid the valley of the shadow of death, or to know that as we call on Him, He will meet us there? We must pray with passion, not for our lives to be "easy," but for the Lord Himself to be with us, our children, and our loved ones no matter what comes.

It's a hard promise for some to hear. David Watson knew it as he lay dying of cancer. He wrote:

> God offers no promise to shield us from the evil of this fallen world. There is no immunity guaranteed from

> sickness, pain, sorrow or death. What he does pledge is his never-failing presence for those who have found him in Christ. Nothing can destroy that. Always he is with us. And, in the long run, that is all we need to know.[4]

Too often, our sights are earthbound. Our focus is for God to shield us and our children from harm, and to provide blessing in this life. That's all we want. We don't want to set our sights on heaven or hear that God doesn't promise "an easy time, just a safe arrival." Why even talk about a "safe arrival"? Because that's what His "never-failing presence...in Christ" is all about. It's the only hope in a troubled world.

A Safe Arrival

Soon after the crash, nearly all the television and radio stations in Pittsburgh began broadcasting the song "As We Sail to Heaven's Shore." It was played at every public memorial service. Some of the words are inscribed on the monument at Sewickley Cemetery. The soothing voice of Kirk Lynn, the gentle flow of melody, and the tender, prayerful words gave comfort and more. It did what songs sometimes do: It touched our grief—as if it knew what we were feeling—and gave it form and expression.

The words of the song portray our lives as a journey aboard a ship; we are yearning with hope for a safe arrival on heaven's shore. It was hard, sometimes, to hear the words. Kirk's voice alone brought the tragedy of Flight 427 physically near, and as a result, made us feel the sadness. Still, he sang with quiet conviction and deep faith—in that voice that was now silenced in death by the crash.

The words of the song became ours. The images of the ship in full sail and heading for port, and the airborne 737 in flight and ready to land, became one and the same. The voyage, far greater than being about the inconvenience of getting from place to place, was about us—the whole of our earthly lives and of the lives of those aboard Flight 427. What would be our final destination? With each passing word, we could hear Kirk's passion rise as he set his focus on the Captain of the ship; his trust was in the Captain's steady hand to guide him safely home to heaven's shore. These words, that voice, weren't asking for an easy life or a smooth sail. Rather, they embraced the end of the journey and made our hearts fill with comfort, soar with hope, and long for that same shore.

> Storms may rise on seas unknown
> while we journey t'ward our home.
> Surely we'll learn what grace is for
> as we sail to heaven's shore.
> Send us strength, O pilgrim guide;
> sin would drown us in its tide.
> Be close at hand, and go before
> as we sail to heaven's shore.*

The song begins with real life—the "storms." On a ship or on a plane, nothing makes the journey more uncomfortable than foul weather. We get battered around, feeling disoriented and confused. It's the same with the other storms that come against us: sickness, a broken marriage, the loss of a job, loneliness, grief. We feel as if we're "on seas unknown." We want the journey to end before "sin would drown us in its tide."

---

*"AS WE SAIL TO HEAVEN'S SHORE" by Phill McHugh and Greg Nelson.

At such times, two issues become all important. One, we must turn to the pilot—the "pilgrim guide" and the Captain of the ship. We need Him to be "close at hand," "to go before," and to give us "strength" and "grace." No one else can make us feel safe. His confidence is our confidence. Two, there's no other prayer than to get to our destination. In this life, we rarely think of "heaven's shore," let alone consider it "our home." Our destination is usually a shortsighted goal: a career move, an addition to the house, or a college education for the kids. But that changes when storms rise and our lives are threatened. This song made us feel the storm and think differently about our journey, the Captain, and our real destination.

> Holy Spirit, lead us on,
> give us courage, bring a song.
> Lord, we trust Your Father's care
> will convey us safely there.
> Open or seal off ev'ry door
> as we sail to heaven's shore.
> Strengthen our course with ev'ry prayer,
> let heaven's breezes speed us there,
> and grant us mercy evermore,
> as we sail to heaven's shore.

By this point, the song has become a prayer. Dependence on the Captain has grown deeper: "Lord, we trust Your Father's care." In the same way, there's trust in His provision for us, now, in the storm; we hear that in the phrases "lead us on," "give us courage," "strengthen our course," and "grant us mercy." But there's a twist. Why not ask the Lord to stop the storm? Why not plead for a peaceful day's sail? Why not pray, "Let heaven's breezes come with force, blowing the storm's fierce winds away"? Why not ask for the "Father's care" to physically save us in the time of trial?

But that's not the prayer of this song. Only one thing

matters, and it's repeated over and over again: the sail to heaven's shore. The breezes from heaven are meant to "speed us there," and the Lord's love is to "convey us safely there." It's about destiny—"the journey t'ward our home." And that, in light of the passengers and crew of Flight 427 and their sudden death, makes sense. What about destiny? What made Kirk sing this prayer eleven days before the crash? Why didn't he pray for a safe trip—a long and happy life on this earth? Instead he asked for and received a better promise—a safe arrival.

> Draw us near, O finest Friend,
> from dawn's light to evening's end.
> Each passing day we love You more,
> as we sail to heaven's shore,
> as we sail to heaven's shore.

By the song's end, we're still aboard the ship, still in the journey. The last prayer is to the Captain: "Draw us near." But this time, it's not for safety from the storm, nor is the Captain someone we don't know. This "finest friend" is the Lord Himself; our relationship to Him is everything, whether we are in time ("from dawn's light to evening's end" and with "Each passing day") or on "heaven's shore." To know the Lord is to have the eternal promise of loving Him more. It is here that prayer reaches its highest peak.

Here is the secret to facing life and to our destiny: To love Him more—not to love Him for what He does for us, but for Him. It is to grow in a deeper relationship—day by day, more and more—with our "finest friend."

I talked to Phill McHugh, who wrote the lyrics to this song made famous by Christian recording artist Steve Green. He said he wrote it in an airport. Phill had spent some time with a close friend who was struggling with personal problems. When their time ended, he put the friend on a plane. But walking back through the airport to his car and saddened by his friend's difficulties, he stopped and watched the crowd.

It was a sea of faces, all in a hurry to go somewhere. But where? Why were so many people dressed in expensive clothes, taking confident strides, appearing so directed, controlled, and successful, as if life was an easy game to play? Bus terminals, he said, aren't like that. All kinds of people go there, but most have less money, less confidence. But for people of every walk of life, from rich to poor, there's one thing in common. We all have an eternal destination. Where were these people going, and how many in this sea of humanity knew their final destination?

Phill said he remembered Jesus looking over Jerusalem and weeping. The Son of God had come from heaven, and most people couldn't see it. They were blind, rushing here and there, busy with their personal lives, completely unaware that He, the Captain, had come to prepare their way to heaven's shore. He had their eternal destiny in mind. He knew the nature of this world, that it is entrenched in sin and full of storms. But they, instead of turning to their Captain, turned away from Him. For this, the Lord Jesus wept (see Luke 13:34-35 and 19:41-44).

Phill felt that weeping, that compassion for the crowd. He saw their mad rush, with tickets in hand and flights in mind. He saw them looking at their watches and speeding to their planes—and for what? Isn't life more than racing through an airport like mice in a maze? That's the sadness—that people are in control but in such a hurry; they know their next airport, but do they know their final destination? Do they realize that one day they'll climb aboard a different kind of plane and head for their eternal home? What then? Where will they go? With love for the people filling his heart, Phill put words on paper. He knew that only one journey in life matters, and that there is only one Captain who can get us safely there. The plane became a ship; the journey takes us not from city to city, but through our whole lives as we're caught in a frenzy, a fierce storm with no hope of survival but to "sail to heaven's shore."

The lyrics were written that day at the airport in 1986. Eight years later, a twenty-six-year-old man stood up in church and sang this song. It's reported that he wanted people to hear the message. So he asked the congregation not to watch him sing, but rather to bow their heads and listen to the words. Then he sang the song. Eleven days later, we stopped our busy lives and heard that young man's voice over the radio and television. One thing was clear: In his death, he showed us what the words meant.

A Changed Life

"But it doesn't last long," a friend from Uganda, East Africa, told me.

"I don't understand."

"When death comes suddenly, people get shocked. They stop their busy lives because they know it might happen to them. Then they make big promises. Churches fill up. Debts get paid. People forgive each other. Husbands become more devoted to their families. Everybody wants to be together. See, we get afraid that our time might be short, so we take stock of our lives and put them in order.

"As a priest, I see this all the time," he went on. "People I've never seen come to church holding Bibles in their hands, sometimes with tears in their eyes. They are scared that death is waiting for them, and they know their lives are not right with God. They're ready to repent of their sins, turn from their wrong, and live for Jesus Christ. It's like an alarm has gone off, warning them that life is short and death is coming, and they've got to make the choice now, while there's time. *Are they going to live for the Lord?"*

"But you're saying it doesn't last."

"For many people, it doesn't. A few months go by, and they forget, as if the warning never came. The shock wears off. They want to return to their old way of life. It's easier not to think about death, or being right with God, or where they'll spend eternity. A confidence comes back that causes them to take risks—to choose things they know are wrong and see if they'll get caught. Why go to church, read the Bible, do what's right, and live for the Lord? It's better to pretend that death will never come near—that there's plenty of time!"

"In their hearts, they know that's not true," I said.

"When I was a child, there was a swamp outside our village. All the children loved to play and swim in the water, especially on hot, dry days. But it was even more fun because our parents told us not to go there. So the kids in our village would sneak down to the swamp, play for a while, and sneak back before our parents knew we were missing. It was all part of the game—not to get caught.

"But there was a reason for our parents to be concerned. The swamp was deceiving. As I walked in the water, my feet were stepping on a grassy netting that was very tight and elastic—like a trampoline. As kids, we loved to jump up and down on the netting, splashing in the water. But it was very dangerous. The netting had gaps. A child would be bouncing up and down and then disappear though the netting."

"Under water?"

"It's just like a person falling through the ice on a frozen lake. Sometimes the kids came straight back up through the hole with no problem. But other times, they didn't because of the undercurrent. On the way up, all they could feel is the grass netting everywhere. And they would panic, looking for the hole and not finding it. It was very scary. The only hope would be that someone saw them go down and that they'd look for the hole, too. If they found it, they'd put their arms through and search for the lost child. Sometimes that worked."

"Children died under that grass netting?"

"Yes. Every few years a child would drown."

"Didn't that scare the kids from going back?"

"Yes, it did for a while. But it didn't last long. Eventually, we'd go back. At first we made up rules for ourselves—swim together, don't stay too long, don't go too far out or not jump so high. But over time, we became brave and broke the rules. We knew what could happen, but we chose to forget. So we swam alone, stayed longer, went out farther, and jumped higher. It became part of the game even though we knew other kids had died there. But you think to yourself, 'It won't ever happen to me or my friends.' "

"Did it?"

"I went down through the netting once."

"You're kidding!"

"Eight or nine of us were jumping up and down, laughing. One time when I came down, I went right through a hole. By the time I realized what happened, I was in the water, under the grass. I started looking for the hole, but it wasn't there and I got scared. I hadn't taken a breath; I didn't have much air. My eyes were wide open. My hands were trying to poke through the netting. But I couldn't find the hole. I did everything I could. And then, when I thought it was all over, I saw these big hands plunge through the grass right in front of me. They grabbed me tightly around my shoulders and yanked me out of the water.

"A man from the village had been spying on us, saw me go down, and saved me. I never went back. I made a choice that day to stop playing with life. I knew it was dangerous to swim there. When I went under the water, it's like I expected it. Why had I been so foolish? Why had I disobeyed my parents? Why didn't I learn from the kids who had died at the swamp? Did I think I was exempt from death somehow? And yet there I was, helpless under the water, with no air and no way out. It was high time I learned my lesson."

"Most of us don't."

"Most of us are afraid to think about death. We're like little kids playing at the swamp, knowing that it's dangerous. We get scared when someone dies. The fear of death changes our lives, but only for a short time. We forget and find ourselves back in the water as if it never happened. I didn't want that. I needed something more than the fear of death—something that would change me forever.

"And that happened when Jesus Christ came into my life a few years later. He showed me that the secret to facing life is found in facing death. He saved me, and deep in my soul I knew that I didn't have to run from death or fear death anymore. No matter when it comes or how it comes, the Lord Jesus has promised me eternal life with Him. And I know what's going to happen. On that day, I'm going to look up, just like I did under the water. And these big hands will plunge down from heaven—right in front of me. They will grab me tight, yank me from death's grip, and bring me safely into His arms of love and glory. I know that. And I live life differently because of it. No more pretending death isn't around. No more playing games with life. I've given myself to the Lord, and I want to live every day for Him."

---

[1]Watson, *Fear No Evil*, 140-141.

[2]Dietrich Bonhoeffer, *The Cost of Discipleship* (New York: Macmillan Publishing Co., 1963), 19-20.

[3]Ibid., 99.

[4]Watson, 159.

Chapter Fourteen

Facing Death

"Papa, what is it like to die?" Corrie Ten Boom asked her father.

"When you and I go to Amsterdam, when do I give you your ticket?"

"Just before we get on the train."

"So it is with death. When the time comes, our wise heavenly Father will give you all the strength you need."[1]

It was such a simple truth: The ticket would be there "when the time comes,"and not before.

On December 13, 1991, Bob Merriman made his annual visit to the doctor. He had no complaints, though many of us had noticed a weight loss—and on Bob, that was hard to tell. Although he was a voracious eater, this short, hardworking man in his early sixties was also very thin. "Call me anything you want," he used to say, "just not late for dinner!" The doctor found a mass that December day, in the upper part of his left lung.

When the steel mills folded in 1982, Bob had been in his fifties and not ready for retirement. He was a gifted carpenter and a skilled fix-it man. Routine calls from friends and neighbors needing his help soon became his full-time work. He made his business official by donning a dark blue baseball cap. In bright yellow letters across the front it read, “Hardly Able Construction Co.” With that, he felt legitimate.

Bob and his wife, Dot, were part of our family at church. In 1988, our Sunday services were held in a fire hall. Construction on our first building began, and Bob—joined by a team of men—designed and built all of the woodwork in the sanctuary. It was a labor of love for the Lord. I volunteered on occasion, although admittedly, I’m not skilled with my hands. Still, I felt I deserved an official business cap, and one day, Bob presented it to me at church. It was white with bright red letters that read, “Able? Hardly!”

We were friends, but we felt more like family.

Christmas of 1991 was hard on us all. Bob was there as if nothing was wrong. Wearing a cheerful smile and quietly whistling, he helped us decorate the church. In his clear, bass voice, he read the Bible at service, with confidence and faith in his Savior. He knew, as we all did, that he had cancer and was dying.

On December 26, he went in for surgery. The doctor removed some tissue from the mass and sent it to the lab. Two days later, the news was grim. It was malignant and inoperable. The only options were chemotherapy and radiation. The fight was on. But Bob knew what was in front of him. Only a few years before, he had watched two close friends die of cancer. It’s not like a massive heart attack, where you’re here one minute, then gone the next. This death takes time. He saw both men, both in full stride only months before, surrender slowly to death. As their strength got weaker, their bodies changed—turning sickly, gaunt, and bone thin. Then came a decline in their mental state; one minute they were lucid, then they’d be confused, incoherent, delirious, and hallucinating. The two dear friends gradually gave up the fight, lost control, and slipped away into silence. What kind of humiliating death is that?

Bob knew what cancer meant. He decided to take the treatments, which began in January. They were physically hard on his body. Near the end of the month, he knelt to receive communion at a service. Erilynne and I prayed for him, and leaning into his wife's embrace, he broke into tears. "I'm not crying for myself. I've always felt the love of Jesus Christ here with all of you. I can feel it in my heart now." We didn't know it then, but that was his last Sunday at church. The downward slide happened quickly.

By early February, the cancer had metastasized to his brain, which meant more radiation. Sitting in his living room later that month, he showed me the four "X" marks drawn on his scalp. "That's where they point the radiation gun and fire," he said laughing. And it was working. The doctors said the four lesions in his brain were getting smaller, as was the cancer in his lung. These were little signs of victory and hope.

His weight loss continued, dropping from 125 to 101. But Bob's mind was sharp and unaffected. Before I left that day in February, I knelt down next to his chair, and we prayed together. At one point he said, "Heavenly Father, not my will be done, but Thine be done" (Luke 22:42). I looked up at him and saw his eyes closed and his head bowed. He had just quoted the same prayer the Lord Jesus Christ had prayed on the night before He died. And Bob did it so peacefully. There were no tears. There was no trembling or begging the Lord for more time. His will to live, to fight the cancer and beat it, was offered back to his Father in heaven in one breath. *Thy will be done.*

I saw it on his face; I felt it as I held his hand in prayer. My friend had been given the ticket.

By mid-March an infection had set in, sending Bob to the hospital. His body was too weak to fight it. He started drifting in and out, his body hot with fever. One minute he was quiet, the next full of happy talk, but nothing coherent. Bob was leaving us. In the third week, he went into a coma. A round-the-clock vigil began with his relatives and church family. In the bed next to him, Jared, age twenty-eight, was also dying of cancer. He and his family knew Christ as their Savior. The two families came together in prayer, both standing watch for Bob and Jared. And one morning, Bob came out of the coma; the infection was gone.

He was home in April. The dining room became his bedroom so that visitors could come and see him. He couldn't walk, nor was he mentally the same. But there was a childlike humor about him. He would talk, with a big smile and those warm eyes, telling story after story. He made it through Easter, singing songs he knew so well—*"Christ the Lord is risen today!"*—but his time was coming near.

On April 27, Dot called us. By half past eight, his family and friends were at his side. He was thrashing for life, his heart at full speed and his breathing labored. Near nine, he began to slow down. Dot bent over and kissed him on the forehead. We prayed, sang a song, and then quietly, Bob took his last breath and died.

Never Alone

What is it like to die?

My life is crowded with little vignettes of death. I've told the story of my mother sending me to school my first day. She hugged me, saying, "Oh, I wish my little boy wouldn't grow up." But the time had come, and she held the door open. I told her, "I don't want to go," and I didn't. Home was safe and known. One step out, it's a new world—unsafe and unknown—with no way back to the old way of life.

I went crying; I didn't like the feeling of loss, and that's what it was. No more days with Mom alone—shopping, cleaning, seeing my grandparents, and taking afternoon naps. No more chances to stay close to her when trouble came near. I was leaving behind a world I trusted—one that was controlled and protected. One look outside, to that new world out there, and I felt scared. I was going alone, on my own. *What if I can't handle it? What if trouble comes near? What if I find myself lost, helpless, unsure? What then?*

Years later on our first trip to the crash site as I stood outside the yellow tape, I felt some of those same feelings. I was all grown up, yet I could still feel the old fire burning. A voice whispered, *Stay safe. Don't go over the tape into that wasteland of death and destruction. You'll lose control. You won't be able to handle it. And what then? Don't surrender yourself to what's unknown. You'll be lost, helpless, alone, and afraid. It's not worth it.* It was the same old warning: *Don't take the step; you don't have the inner strength.*

And I didn't.

With each glimpse of death, I experience a deeper surrender of myself. First, it's losing "safe" in my life, where I have control, purpose, love, connection, and security. Second, it's stepping into a new world that promises no control, no connections. That's when the fear begins. When I die, will I lose myself? Will I lose those I love? Will my family and friends watch my body slip slowly away in death? Will they see me disconnect from them and from myself, from our history and our love? Will there be any sense of dignity in dying? Will I be lying in my bed, surrendered, without control, and without knowing?

Will I be alone? Will I die like someone out in the cold—helpless to fight the elements, far from home, surrounded by those I don't know, and hungry for those I love? What would it be like to be disconnected from family and from God, completely alone in death for all eternity? Is there any greater fear than this?

Only those who have seen glimpses of death know this fear. Sometimes it's the little things that bring it to the surface, like a new ache in my body (*What is it?*), an odd look in our child's eyes (*What if she weren't here?*). Perhaps it's a turbulent ride on a plane, a car accident in the other lane, a house fire in the news, a rape just down the street, a crack of lightning overhead, or the deep rumble of thunder that shakes the soul. That's when the glimpses come, and we know that there's a door, a yellow tape, and a time that will come for us. We'll have to take that step out, lose our safe place, surrender our control, and face death. How are we going to do that?

I have seen people die alone, and I've never understood it. I've told them that there's a ticket that was bought at a great price. I've told them they don't have to be alone—now or ever—and that that's the only way to face death. I've seen people refuse that ticket despite my pleading, as if they've never felt the fear of God, of dying, of being alone. How is that possible in this fallen world, which daily demands that we consider our death and eternal destiny?

As for me, I know one thing for certain: I will get that ticket. And in death, I will not be alone.

How do I know that? Two thousand years ago, God came to be with us. Jesus Christ, the eternal Son of the Father, left His home in heaven—safe and known—and "emptied Himself, taking the form of a bond-servant" (Phil. 2:7). He was born into our humanity and faced the dark powers of the world, the flesh, and the Devil. God came as man, and He should have been honored, followed, and obeyed. Instead, "He was despised and forsaken of men, A man of sorrows and acquainted with grief; And like one from whom men hide their face, He was despised, and we did not esteem Him" (Is. 53:3). Jesus knows what it's like. He took the step, left His home, and entered a world that despised Him, mocked Him, spit on Him, stripped Him, whipped Him, scourged Him, and crucified Him on a cross.

He did not deserve to die. Sin has always led to death, and the Bible says that He was sinless (see Gen. 2:16-17; Rom. 6:23; Heb. 4:15; 2 Cor. 5:21). On the night before His crucifixion, Jesus prayed, "not My will, but Thine be done." In that moment, He surrendered Himself to His Father and to the plan made from the beginning of time. He came to take our sins and to die in our place; "He was pierced through for our transgressions, He was crushed for our iniquities" (Is. 53:5). It's beyond our understanding and too great to imagine, but our God has faced death. He has "humbled Himself by becoming obedient to the point of death, even death on a cross" (Phil. 2:8).

And it was real death. The nails tore through His flesh into the wood. The cross was lifted up into the air, leaving Jesus alone to die. The crowd jeered Him. Why, if He saved others, didn't He save Himself? Was He now helpless—with no power? The soldiers stripped off His clothing and gambled for it, leaving His body beaten and bruised, exposed in humiliation for all to see. Was there no dignity? But how could there be on a cross, where only the cursed were sent to die? Alone—completely alone—Jesus died. Many of His followers had deserted Him—even His closest friends. And worst of all, so did His Father.

Jesus cried, "My God, My God, why hast Thou forsaken Me?" (Matt. 27:46). How could His Father have forsaken Him in death? But that, too, was part of the plan. God had placed on His Son the sin of the world. There on the cross, Jesus bore the punishment for our wrong. He was dying our death. He "who knew no sin" became "sin on our behalf" (2 Cor. 5:21). It cost Him everything. It meant that the Father, in punishing sin, had to turn His back on His Son. It was the only way. The cross had to happen. Jesus had to die a real death, alone and God-forsaken, so that His blood could be shed for the forgiveness of sin for all who believe in Him.

There is no reason for anyone, ever again, to die alone.

God Himself has come in Jesus Christ and faced death. He knows what it's like to step blindly and helplessly into the unknown. The Lord God, holy and pure, had never experienced sin, darkness, evil, injustice, loss of control, loss of dignity, abandonment, or death. And He did. He experienced it. He surrendered Himself on Calvary for us. The Lord Jesus Christ knows death, and therefore, He knows how to answer our cries when death comes for us. We don't have to fear being forsaken; He did that for us. Nor should we fear losing our safe place, or losing ourselves, or being left out in the cold with no strength and no one beside us in death.

In Jesus Christ, death has lost its power, its sting. We can opt to die alone, but why do it when we can turn to Him and see in His nail-scarred hand—clutching it tightly just for us—the ticket home?

Triumph!

Before he died of cancer, the priest and evangelist David Watson wrote these final words:

> But for those who know God and who are trusting in Christ as their Saviour and Lord, there is nothing to fear, and it is sufficient to know that we shall be like him and perfectly with him. Nothing could be more wonderful than that. Never fear the worst. *The best is yet to be.*
>
> When I die, it is my firm conviction that I shall be more alive than ever, experiencing the full reality of all that God has prepared for us in Christ. Sometimes I have foretastes of that reality, when the sense of God's presence is especially vivid. Although such moments are comparatively rare they whet my appetite for much more. The actual moment of dying is still shrouded in mystery, but as I keep my eyes on Jesus I am not afraid. Jesus has already been through death for us, and will be with us when we walk through it ourselves.[2]

How can a man on his deathbed write such words? One day and one message lie at the very heart of the Christian faith. It's Easter day and the message that Jesus Christ has risen from the dead. "Do not be afraid," the angel of the Lord said to the women coming to His tomb, "for I know that you are looking for Jesus who has been crucified. He is not here, for He has risen, just as He said. Come, see the place where He was lying. And go quickly and tell His disciples that He has risen from the dead" (Matt. 28:5-7).

The tomb was empty, His body gone. Many people all over the world believe in life after death and that the spirit lives on. They have no proof, but it sounds right and comforting. That's not what happened on the first Easter morning. The Christian faith begins with an empty tomb. Jesus *bodily* arose from the dead—not just in spirit, but physically. He stood in front of His own disciples after His death and said, "Why are you troubled, and why do doubts arise in your hearts? See My hands and My feet, that it is I Myself; touch Me and see, for a spirit does not have flesh and bones as you see that I have" (Luke 24:38-39).

Jesus is alive! He physically arose from death.

His resurrection has one meaning: He defeated death. Everyone else—even the greatest men and women the world has ever known—has died, and their bodies have turned to dust. Death isn't partial to anyone. All of us have sinned. All of us deserve death. But not Jesus of Nazareth. He went to the cross without sin. He alone, the Son of God, was born perfect, just as God intended us to be. At Calvary, He took our sin to Himself—not any of His own. When He died our death in agony and pain, having paid in full the penalty for our crimes, His work was finished. "And God raised Him up again, putting an end to the agony of death, since it was impossible for Him to be held in its power" (Acts 2:24). For the first time ever, death lost!

Easter is nothing less than victory over death. Christians are "Easter people." We celebrate the resurrection of Jesus because He has conquered death on our behalf: " 'Death is swallowed up in victory. O death, where is your victory? O death, where is your sting?' The sting of death is sin, and the power of sin is the law; but thanks be to God, who gives us the victory through our Lord Jesus Christ" (1 Cor. 15:54-57).

That victory is for all who believe in Jesus. He Himself said, "I am the resurrection and the life; he who believes in Me shall live even if he dies" (John 11:25). But we must believe in Him. We can choose to live our lives without Him, doing our own thing, going our own way. And when death comes near, no matter how or when, we can die alone, unsure of our destiny. But why do that? Why gamble? Why do we live as if we don't need the Lord and as if death isn't near? Why don't we turn from our sin and stop doing the things that displease Him? Why don't we live for Jesus Christ now, while there's time?

What does it mean to be a Christian? First, it means we've faced our true condition. We are born into the world as sinners—without God and bent on doing wrong—and there's nothing we can do about it in our own strength and power—no acts of charity, no long prayers, and no years of attending church. There are no bargains with the Almighty. We are guilty, and nothing we offer can make us right with God. We are responsible for our sin and wrong actions. Without forgiveness, we face death and eternal separation from the Lord.

A Christian knows the seriousness of sin and holds fast to these words: "For God so loved the world, that He gave His only begotten Son, that whoever believes in Him should not perish, but have eternal life" (John 3:16). At first, the news is hard to hear; we deserve to "perish," and if we die without God, that is what we must face. But the good news is this: God loves us. He doesn't want us to perish in our sin. He has sent His Son Jesus Christ that we, believing in Him, might have "eternal life."

Second, being a Christian means we've confessed our sin with repentance. It isn't enough to say we're sorry and not change. Apologies never work without a choice to stop the behavior. We know we have sinned and that we need God's forgiveness. In just the same way, we know our lives must change.

Third, a Christian has experienced the grace of God.

Something has happened deep in our souls, opening our hearts and minds and showing us that Jesus Christ died for our sins. His blood was shed for our forgiveness. He is alive today, risen from the dead. The tomb is empty! He is the Savior, and we need Him to save us. Sometimes this understanding comes while hearing a sermon, talking to a friend, or reading the Bible. But God begins to open our eyes. We see the gravity of our sin and our need for Jesus Christ to give us new life.

Finally, a Christian receives Jesus Christ into his or her heart. He alone can rescue us from perishing and from our natural inclination to do wrong. For years I attended a church where no one told me I must turn from my sins and ask Jesus into my life. But one day in school, a friend led me in prayer to receive Jesus Christ as my Savior and Lord, and my life changed. From then on I knew I belonged to Jesus, and I wanted to live for Him. By God's love and grace, my heart and mind were opened, and I became a Christian.

How could David Watson pray on his deathbed without fear? He knew his Savior. He wrote, "But for those who know God and who are trusting in Christ as their Saviour and Lord, there is nothing to fear." He knew the secret. Jesus had defeated death. It has no power for those who trust in Him and His victory over the grave. So Watson wrote, "I am not afraid. Jesus has already been through death for us, and will be with us when we walk through it ourselves."[3] David faced death with the Lord Jesus Christ.

There is a ticket. There is an answer to the question of what it will be like to die. But we must believe in the Lord Jesus Christ. Why live without Him? Why think that our lives can be free from death, as if the twenty-three seconds could never happen to us? I think now is the time for people to give their lives to Christ and live for Him. Have you acknowledged yourself as a sinner? Have you confessed your sin to the Lord and repented by turning from your wrongdoing? Have you asked the Lord to open your heart and mind so that you might know there is a Savior who died and rose again for you? Have you ever received Jesus Christ as your personal Lord and Savior? Today, while there's still time, give yourself to Him. Invite Him into your life. Find a quiet place, think about these things, and take the step by praying this prayer:

> Lord Jesus Christ, I acknowledge that I have gone my own way. I have sinned in thought, word and deed. I am sorry for my sins. I turn from them in repentance. I believe that you died for me, bearing my sins in your own body. I thank you for your great love. Now I open the door. Come in, Lord Jesus. Come in as my Savior, and cleanse me. Come in as my Lord, and take control of me. And I will serve as you give me strength, all my life. Amen.[4]

It is a simple prayer, but it changes everything. A new relationship has begun. Each day as we spend time in prayer, reading the Bible, and being with other Christians at church, our relationship with the Lord grows and deepens. In Him we learn how to live in this life, and when the time comes, we learn how to die. In everything, Jesus has made this promise: "I am with you always, even to the end of the age" (Matt. 28:20).

Relationship! Our bookstores, talk shows, and business seminars are full of advice on how to master life and death. Other religions have their philosophies and teachings. But what can compare to this? God has come in Jesus Christ. He has provided a way for us to be in a relationship with Him. And nothing, by His own decree, can ever come between Himself and a person who has believed in Christ. Nothing!

In this life, suffering will come, along with sickness, pain, tragedy, accidents, earthquakes, fires, thefts, rapes, murders, and injustice to our rights and the rights of others. Sometimes death will be sudden, sometimes slow. But all the suffering in the world cannot break the bond of this relationship, which was forged at the cross and won at the empty tomb. No matter what life does to us, a Christian can rise in the morning without fear, for the Lord is with us. He will keep us safe to the end and give us the strength we need to get through it all.

Early Christians in Rome died for their faith in Jesus. The apostle Paul needed to tell them the same truth—that nothing can separate this relationship. The love of God, in Christ, to the believer is forever. It can face death. It can face life. It can face anything. There is triumph in belonging to Jesus:

> For I am convinced that neither death, nor life, nor angels, nor principalities, nor things present, nor things to come, nor powers, nor height, nor depth, nor any other created thing, shall be able to separate us from the love of God, which is in Christ Jesus our Lord (Rom. 8:38-39).

He's All I've Got

I didn't see Jared at first. His hospital bed was to the right of Bob Merriman's, and the curtain was drawn. I was struck by his appearance when, that night, his family came and pulled the curtain back. His eyes were huge and beautiful, almost otherworldly, set to perfection above high cheek bones and a strong, chiseled face made thin by the cancer. His unlined forehead sloped gradually back to a bald head, and his young, cocoa-brown skin stretched tightly over each bone. Both his legs were together under the sheets; they extended straight to the end of the bed. His right arm—the arm closest to us—lay still on his lap. His other arm was gone.

Jared was sitting up, hunched over with his face down, and fighting for breath. When someone talked, he looked up, but it was uncomfortable and tiring. It was easier to breathe bent over—and easier not to talk. In those few days at his bedside, I never saw him lie down, nor did I hear him groan with complaint. Cancer had started in his arm; the amputation wasn't fast enough, and it spread to his lung and collapsed it. The treatments couldn't stop it, and there was nothing left to do but to keep Jared as comfortable as possible.

His mother said it was hard for Jared to accept at first. He had done everything right—no smoking, no drinking, no hanging with the wrong crowd. He was a Christian. He had made his choice to live for Jesus and follow Him. He had joined the service, married a sharp, attractive nurse and had two children. I saw them—a five-year-old girl who was blessed with his eyes, and an eleven-month-old baby. "But it's hard," his mother told me, "when you're twenty-eight. You want to live, see your children grow, and make a difference in other people's lives. And that's not going to happen. He knows that now. We all do."

For three nights, his side of the room was full of family—grandparents, aunts, uncles, nephews, nieces, parents, brothers, sisters, and friends. He sat in his bed, arched over, and listened to all their stories of years gone by. Each story conveyed with just a few words a memory that linked them together. And each story had the same ending: God had always been with them, blessing them in their joy, and comforting them in their griefs. He was with them now, ever faithful, even in the cancer ward of that hospital. Then, out loud, they thanked Him—in Jared's hearing and in mine.

One late afternoon, he was alone and the phone rang. I went over to get it. "It's your wife," I told him.

He shook his head slowly; he couldn't take the phone. It was just too hard to do anything but breathe.

"She wants you to know she loves you. She'll be in about quarter to seven." He nodded, and I told her so. As I put the phone down, he held out his one good hand, and I took it. He whispered, "That was nice of you," squeezed my hand and let go. It was the only time we had ever been together alone.

"There's something about you, Jared. You've got the most wonderful family and such beautiful children. They love you so much. And I know it's hard to breathe, but you're really at peace, aren't you? I know Jesus Christ is with you." He lifted his head, looked me in the eyes, and said, "He's all I've got." And I knew what he meant. He was ready for his journey, and there was nothing here to take with him—not his wife, children, family, friends, or things. All these would be left behind. Jesus was all that Jared had.

At that point Bob took a turn for the worse, and we began our vigil around the clock. Late one night, it seemed that Bob's death was near. He'd breathe, then pause—breathe again, then pause. Near midnight, another shift came in. His wife, Dot, stayed the whole time, but she opted to take the shift at three by herself. The hospital room was dark, Jared's curtain was drawn, and Bob lay quietly in his bed, still close to death.

Dot sat in a chair between the two men.

Just after three, she leaned on Bob's bed and stared into his face. After the four-month fight against cancer, here it was—the inevitable. Her husband of forty-two years, who had pledged to be at her side "until death do us part" was leaving her. Her mind began to roll out memory after memory—their first date, their first child, vacations at Rehoboth beach, Bob teaching his grandsons to hunt, drive, and work with their hands. How could she live without him? Late in the night with no one around, Dot's tears turned to sobs. Out it came, the crying she'd pent up for so long so that she could be strong for Bob. Now there was no more strength, no more fight—not in Bob and not in her. She was weeping, her heart empty and alone.

She sat back in her chair, praying for her husband and for comfort, then drying her eyes.

And then she heard a sound. In the silence of the room, there was the sound of metal on metal—a movement. She knew the sound; it was the curtain behind her, the rings up above sliding on the metal rod. She didn't move; then a hand came to rest on her left shoulder. It patted her and gave her comfort in her sorrow.

Jared.

She put her hand on top of his—that one good hand—and it made her cry all the more. Who was this man helping her in her sorrow and bringing such love to her? How could he do that in his suffering? How could he reach out—with just the right touch—in his dying? He squeezed gently, tenderly, keeping his hand there, firm, with such affection. He was like a son consoling his grieved mother. Then his hand lifted. She turned to look at him; he was still bent over, still working for each breath. In the dark, she could see his eyes looking at her. He didn't say anything, but she knew, as if he said aloud that it would be okay.

"He's all we've got."

Morning came, and Bob was stronger. His breathing was back to normal; the fever was gone. The doctors were surprised. By evening he was taking liquids, talking, and recognizing his visitors. Jared's family came in. As the night wore on, they pulled the curtain between us. I didn't think much of it; they wanted their privacy, that's all. I'd gone home by the time the curtain opened again. My friend Nick Petkovich told me that when the curtain opened, Jared was lying down in bed, his body still. There would be no more gasps of air or sleepless nights. The watch was over.

"Tell Me, Why Did This Happen?"

Why Jared? He was twenty-eight. He did everything right—lived a good, clean life.

Some people see God through the lens of their grief. In their eyes, God is a monster—unfair and unkind. They ask why God would inflict a guy like Jared with cancer. Or if He didn't do it, why didn't He step in and heal him? He has the power to heal, change circumstances, and protect from evil and death. Didn't He love Jared, wanting with all His heart to act on his behalf? So why didn't He do it? Why didn't He answer the prayers of a five-year-old for her father or of a wife for her husband? What kind of God is this?

Grief is blinding. There are no answers to their questions—not in their condition. No answer will satisfy the hearts of those who are mourning. They want one thing: their dead back. And that's not going to happen. So they choose to see God through their grief, blame Him for their loss, and hold Him in contempt for His lack of care.

Other people see their grief through the lens of God. For them, a whole new horizon opens. Jared's family grieved for his death, but they knew about this broken, fallen world full of sickness and death. God didn't make the world this way. He didn't cause Jared's cancer. Sin came into the world, not by God's doing, but by ours. And when sin came, so did suffering, sorrow, and death. It is part of this life, and it will be until the end. Jared's family knew that. His death saddened them, but they understood it.

They knew what Jesus Christ had done for him. He did step in, with power and love, acting on Jared's behalf. He did far more than deal with the cancer. He dealt with the sin. The Lord Jesus saved Jared, not for a longer earthly life, but more, far more—for eternal life. That's why Jared's family could give thanks to the Lord in the cancer ward of a hospital. The Lord was with them, faithful and true. The cancer would destroy Jared's body, but not his soul. In his death, he'd be "carried away by the angels" (Luke 16:22) into the presence of the Lord. Nothing could take that from him. It was promised him on Calvary.

We have a choice to make. We can see God through our grief and never understand why there's suffering in this world, or we can see our grief through God's eyes. If we do, we'll see the promise of heaven and the glory that awaits all who have put their trust in Jesus. "For I consider," wrote the apostle Paul, "that the sufferings of this present time are not worthy to be compared with the glory that is to be revealed to us" (Rom.

8:18). In all the sufferings we face, nothing can compare with the glory of heaven to come.

"Tell me, why did this happen?"

I had to answer that question on live television. We had just come from the crash site on the first day. It was hot news: Flight 427 had lost control, spiraled to the earth, and killed 132 people in twenty-three seconds. Now the reporter wanted the clergyman to answer why. He put the microphone close as the camera zoomed in on my face. I had only a few seconds to answer. I needed to be clear. I had something to say. There is an answer to that question. I looked into the camera and said:

> I know it's a shock to everyone. But, let me say this: It isn't the first plane crash. We live in a fallen world where things like this happen all the time—accidents, sickness, fires, and death. God doesn't cause these things to happen. He never promised life here on earth would be easy. It wasn't easy for Him—He was nailed to a cross and left to die. But for everyone who calls on His name, He promises to be with them in their suffering. And for everyone who believes in Jesus Christ, He promises eternal life. We must not blame Him for this crash. Now is the time to turn to Him for comfort and trust Him more than ever.

It's a choice we make in suffering. We can turn against the Lord and walk away, or we can turn to Him and find that He knows all about our pain. There are marks in His hands, His feet, and His side to prove it.

What is it like to die?

I suggest this: If we'd choose to see through God's eyes, then we'd know how to handle the storms in this life. The storms would come, trying to knock us off course, but we wouldn't fear. Our hearts would be set first on Jesus, who loved us and died for us. No matter how fierce the storm, He has promised a safe arrival into His eternal Kingdom for all who believe. He has the ticket—and when it comes time to die, it will be there.

He will be there. But we must believe.

At 7:03 on Thursday night, September 8, 1994, a plane went over the Hopewell soccer field. A mother stood on the sidelines with her four-year-old son at her side. She was watching her daughter practice and paid no attention to the plane overhead. A few days later, her story was heard everywhere.

The young boy at her side never took his eyes off the plane. Seconds before the explosion, her son cried out with his arm in the air and his finger pointing to the sky.

"Mommy, Mommy, look! Look! The angels! The angels!"

She turned, looked into the sky, heard the explosion, and saw only black smoke.

It was the child who saw it through God's eyes. Flight 427 was speeding at three hundred miles an hour straight toward the ground, and the boy saw angels. There was nothing to fear. Even in those twenty-three dark seconds of terror, something else was happening. Angels had come—like God's breezes to unfurl the sails of His own, filling them with strength and power for a new journey, the last journey home.

It's the promise of the Savior for all who have made the choice to set sail to heaven's shore: a safe arrival.

---

1. ten Boom, *The Hiding Place*, 29 and last page of photographs.
2. Watson, *Fear No Evil*, 168.
3. Ibid.
4. John R. W. Stott, *Basic Christianity* (London: InterVarsity Press, 1958), 129.